**CAREERS** in

HORTICULTURE AND BOTANY

Career and Academic
Advising Center

Professional Careers Series

CAREERS in
HORTICULTURE AND BOTANY

JERRY GARNER

SECOND EDITION

New York Chicago San Francisco Lisbon London Madrid Mexico City
Milan New Delhi San Juan Seoul Singapore Sydney Toronto

Library of Congress Cataloging-in-Publication Data

Garner, Jerry.
    Careers in horticulture & botany / Jerry Garner.—2nd ed.
        p.   cm.
    Rev. ed. of: Careers in horticulture and botany. c1997.
    Includes bibliographical references (p.   ).
    ISBN 0-07-146773-4 (alk. paper)
     1. Botany—Vocational guidance.   2. Horticulture—Vocational guidance.
  I. Title. II. Title: Careers in horticulture and botany.

  QK50.5.G37   2006
  580.23—dc22
                                            2006009583

1 2 3 4 5 6 7 8 9 0   DOC/DOC   0 9 8 7 6

ISBN-13: 978-0-07-146773-5
ISBN-10: 0-07-146773-4

McGraw-Hill books are available at special quantity discounts to use as premiums and sales promotions, or for use in corporate training programs. For more information, please write to the Director of Special Sales, Professional Publishing, McGraw-Hill, Two Penn Plaza, New York, NY 10121-2298. Or contact your local bookstore.

This book is printed on acid-free paper.

CONTENTS

CHAPTER 9
**Mentoring Tips from Horticulture
and Botany Professionals**   **139**

> Lisa: Plant science technician • Norman: Plant geneticist and
> associate professor • Helene: Plant pathologist, associate
> professor, and extension faculty member

APPENDIX A
**Universities and Colleges in the United States
and Canada Offering Degrees In Agriculture**   **143**

APPENDIX B
**Professional Associations in
Horticulture and Botany**   **153**

APPENDIX C
Suggested Reading   **163**

ACKNOWLEDGMENTS

I would like to thank the many people I have worked with in the fields of botany and horticulture throughout my career for their inspiration in writing this book. The initial concept for this book grew out of their enthusiasm for their work and their genuine desire to encourage others to explore these fields as career options.

I would like to acknowledge the encouragement I received from my daughter, Lauren, and her numerous contributions to this book. Her academic and professional experience in the fields of botany and horticulture provided me with a unique insight into both fields.

Special thanks go to my wife, Gerry, for encouraging me to write this book, and for all her love, patience, encouragement, and support in bringing it to completion.

INTRODUCTION

ABOUT THIS BOOK

Plants are central to the ecological system of our planet as well as the aesthetic well-being of humans. Consequently, the study and use of plant material have been important from the beginning of time. As food crops, plants have sustained life. As building materials, plants have protected people from the elements. As medicine, plants have soothed and cured many ailments that have afflicted humans.

To most people, the term plants can mean many different kinds of living organisms, ranging in size and complexity from microscopic bacteria to giant trees such as redwoods. Most laypeople include algae, fungi, mosses, ferns, cone-bearing plants, and flowering plants in the definition. Today, most scientists place bacteria, algae, and fungi in their own distinct kingdoms, but many university and college plant science departments still include these groups in their courses.

In the 20th century, basic research and information about plant life increased so significantly that the results of this knowledge have been applied to more corners of our life. In addition, new career areas have evolved that offer opportunities for those interested in plants to pursue careers in a wider range of disciplines. For example, advances in plant genetics have led to new fields of biotechnology in horticulture and forestry. Advances in plant pathology, the study of plant-related diseases, have resulted in new fields of disease management and control in greenhouses, nurseries, and forests. It is no wonder that the study of plants, from the cellular level to the entire plant, provides such a wide variety of career options.

To help you understand the many career possibilities now available in botany and horticulture, this book is designed to

- Provide an overview of the career options in botany and horticulture
- Define the occupational areas in botany and horticulture
- Answer questions about pursuing various career options in the fields of botany and horticulture

Some of the questions this book is designed to answer are

- What are the differences between botany and horticulture?
- What are the occupational options that I can pursue in either botany or horticulture?
- How do I know which area(s) of botany and horticulture is right for me?
- Where are people in botany and horticulture employed?
- What are the earning capabilities of people in botany and horticulture?
- How should I prepare for a career in botany or horticulture?
- Where can I study botany and horticulture?

This book concentrates on career opportunities in the fields of botany and horticulture, as well as options in some fields that are not easily categorized as botany or horticulture careers because their knowledge base is primarily botany but their application is primarily horticulture.

AN OVERVIEW OF BOTANY AND HORTICULTURE

To begin, let's examine the differences between the fields of botany and horticulture.

Botany

The term botany is derived from the Greek *botane*, meaning plant or herb. Botany and its many branches are pure sciences dealing with the fundamental principles of plant life. In general, botanists study the overall development and life processes of plants. These include plant physiology, heredity, distribution, anatomy, and morphology; the environment; and the economic value of plants for application in other fields. The results of the basic research done by botanists are applied to four major fields of study: horticulture, agronomy, forestry, and pharmacology.

Botanists study the chemical composition of plants and understand complex chemical combinations and reactions involved in metabolism, reproduction, growth, and heredity. They also study the behavior of plant chromosomes and reproduction, internal and external structures, and the mechanics and biochemistry of plants and plant cells. They also may investigate plant environments and plant communities and the effects of rainfall, temperature, climate, and soil on plant growth, development, and evolution. Some botanists identify and classify plants, while some conduct environmental studies and prepare reports. Most often, botanists are designated according to their field of specialization.

The results of botanical study can improve the quality of our medicines, food products, fibers, and building materials. These results may also help to improve our management of parks, wilderness areas, forests, or wetlands. An understanding of plant science may even lead to better solutions to problems caused by overpopulation and pollution.

Botanists are not necessarily concerned with applying their ideas outside the realm of research in the laboratory or field. They may be most concerned with understanding the basic principles and laws governing plant structure, biochemistry, and function. Other branches of plant science, such as horticulture, might then apply these fundamentals to the production and utilization of plants.

Horticulture

By definition, horticulture is "the intensive cultivation of plants." The name is derived from the Latin terms *hortus* and *cultura*, meaning "garden cultivation." Today, the term horticulture has evolved to refer to the art and science of growing and utilizing fruits, vegetables, flowers, and ornamental plants for the benefit of people and the environment.

Horticulture itself is the study of the cultivation of plants. Within that field, agronomy is the study of the practical use of plant and soil sciences to increase the yield of crops, forestry is the study of forest management for conservation and production of timber, and pharmacology is the study of medicinal actions of substances.

The intensity of crop cultivation distinguishes horticulture from most other agronomic sciences involved in the production of crops, such as tobacco, rice, field corn, timber, and others. Horticultural crops such as roses must be regularly watered, pruned, fertilized, and sprayed for damaging pests. Loss of a few plants during production could result in economic loss to the grower. However, a corn crop, for example, is dependent on the weather for watering, and requires minimum fertilization prior to planting, as well as minimum spraying for pests—and then only if a major reduction in yield is imminent. In this case, the loss of a few corn plants would not result in any real economic loss to the grower.

Another distinction between horticultural crops and other agronomic crops is the intended use of the crop. For example, pine trees grown as a horticultural crop are carefully pruned, fertilized, and monitored to produce an aesthetically pleasing tree, each tree bringing a high price in the nursery and landscape market. However, pine trees grown for timber need not be aesthetically pleasing, thus their cultivation requirements are less demanding, and each individual tree would not command as high a price for the grower.

Typically, horticulturists are concerned with the study of so-called higher plants rather than lower plants. Higher plants are those that reproduce from seeds produced by flowers or cones, such as petunias and pines. Lower plants reproduce using spores or by breaking apart into pieces that then grow into another plant, such as ferns or mosses. There are exceptions to this distinction, however. For example, ferns are typically grown in greenhouses and in perennial plant nurseries, both of which are the workplace of the horticulturist. Horticulturists most often work with entire plants rather than plant cells, and they are more concerned with plant production than with the fundamental cellular composition, biochemistry, and classification of plants.

Horticulture is an applied science, relying upon many other disciplines such as chemistry, physics, engineering, art, meteorology, economics, entomology, botany, and many more. A horticulturist uses the basic information and theories of these related fields and tries to find an application for these ideas for the benefit of people and the environment.

Botany and horticulture are interrelated but distinct branches of science. Both involve the scientific study of plants, but on different levels and for different purposes. This is not to imply that one career field is more difficult or more important than the other. Both provide a range of opportunities requiring various levels of education and offering various levels of compensation. Because both fields are so broad, each has a wide variety of career choices with an equally wide variety of career paths a person could follow to reach his or her career goals.

As you can see, the terms botanist and horticulturist are very general career titles. Rarely will you find a person in either of these fields who has not specialized somewhat and narrowed down his or her work into one of the subdivisions of these two careers. These terms are simply umbrella titles under which many diverse and more specialized careers can be found. The following chapters thus are meant to provide general overviews of the various types of employment that exist within botany and horticulture careers.

2

CAREERS IN HORTICULTURE RESEARCH

Today, there is increasing emphasis on the quality of the environment, on efficient food production for world populations, and on quality food for better health. These phenomena have created a demand for improved horticultural products and thus a new and expanded level of interest in careers related to plant cultivation and utilization.

Horticulture is one of the many divisions of the agricultural sciences. Agricultural science can include other career fields such as agronomy, forestry, soil science, plant pathology, and more. There are five major branches of horticulture: pomology—fruit production; olericulture—vegetable production; floriculture—flower production; nursery culture—nursery crop production; and landscaping—the design, construction, and maintenance of landscapes. Each of these divisions can be expanded further into different career fields such as the production and marketing of plant seed, nursery, and greenhouse crops, and Christmas trees and turf. Other fields may include private and public gardening, wholesale and retail florists, education, research, design, advertising, photography, and food processing and storage. Even a career in a support industry such as the manufacture and sale of equipment, machinery, structures, pesticides, and other chemicals could result from an interest and training in horticulture.

In an attempt to understand the place of horticulture in the plant sciences, the following list defines the various fields of plant science and their relationships to one another.

Plant Science

- *Agronomy* and *range science* deal primarily with the cultivation of field crops and range and pasture plants.
- *Forestry* concerns the commercial production and utilization of timber.
  - *Silviculture* is the practice of raising forests.
  - *Urban forestry* is the management of trees in urban areas on larger than an individual basis.
- *Horticulture* concerns plants that are intensively grown for food and aesthetics.
  - *Pomology* is the cultivation of perennial fruiting plants, primarily woody trees and vines.
  - *Vegetable crops (olericulture)* is the growing of herbaceous plants for human consumption.
  - *Environmental horticulture* is the cultivation of plants to enhance our surroundings.
  - *Floriculture* is the production of cut flowers and potted plants.
  - *Nursery production (ornamental horticulture)* is the production of primarily woody plants for landscape plantings and fruit production.
- *Landscape horticulture* is the care of plants in the landscape.
  - *Arboriculture* concerns the cultivation of woody plants, particularly trees.
  - *Landscape construction* involves the installation of structural and plant materials according to a landscape plan.
  - *Landscape maintenance (gardening or grounds maintenance)* specializes in the planting and care of a wide variety of plants used in the landscape.
  - *Turfgrass culture* concerns the growing of turf for landscape and sports use.
  - *Landscape architecture* concerns the planning and design of outdoor space for human use and enjoyment.
- *Park management* concerns the total responsibility of planning, developing, and managing public and private landscaped areas, from housing developments and city parks to heavily used national parks.

(The preceding was excerpted from *Plants in the Landscape* (2d edition) by Phillip L. Carpenter and Theodore D. Walker.)

In this chapter, we will examine the more scientific or research-related fields of horticulture, and discuss the career areas of pomologist, floriculturist, greenhouse grower, greenhouse manager, and olericulturist in detail. Jobs such as groundskeeper, landscape designer and architect, floral designer, and horticultural therapist will be discussed in Chapter 4.

WORK SETTINGS

People who make their living as horticulturists do so in a wide variety of ways and in a wide variety of work settings. Horticulturists work in laboratories, greenhouses, orchards, vegetable fields, gardens and parks, unexplored jungles, at drawing tables, and in classrooms. They may work with a microscope, a tractor, a computer, drawing tools, or a pruning knife. Such diversity of tasks and settings means that the field of horticulture can accommodate a wide range of people and interests!

EMPLOYMENT OUTLOOK

With increasing consumer interest in the environment and the efficient growth, production, and storage of food and ornamental plants, the demand for people in horticulture careers will continue to increase. While efficiency in growing plant material will definitely decrease the demand for workers who plant and grow various crops, new areas of expertise will continue to emerge.

Horticulture is a very diverse area with career opportunities for almost any person interested in the field. Jobs exist for almost everyone, from those with no formal education in horticulture to Ph.D. scientists and educators. Opportunities will continue to increase through 2014, especially for those with specific skills and at least a bachelor's degree level of education. The following sections describe specific careers within the field of horticulture.

ADVANCEMENT OPPORTUNITIES

As their experience and education levels increase, horticulturists may advance to positions with more responsibility. For example, a greenhouse technician may advance to greenhouse manager or head grower. A faculty member may be become head of a research unit or a department head. Paths for career advancement are determined by the individual career chosen and are usually limited only by the ambitions, education level, and skills of the person pursuing the career. The nature of many of the career fields in horticulture allows horticulturists to open their own businesses.

PREPARING FOR A CAREER IN HORTICULTURE

How much education is required in the field of horticulture? Some successful horticulturists have less than two years of college, while others hold associate, bachelor's, master's, and even doctoral degrees. The amount of education that you will need depends on the type of horticulture occupation that interests you and the level to which you wish to advance. For example, if you plan to work as a greenhouse grower, you might need a bachelor's degree in botany. If you aspire to nursery management, business skills and perhaps a master's degree will be beneficial.

HIGH SCHOOL PREPARATION

Your high school curriculum will depend on your particular area of interest, but certain courses are recommended for all students. High school courses should include biology, math, English, general business, and computer skills. More specific courses such as plant science, horticulture, floral design, and art would be beneficial depending on your interest.

If you intend to obtain a college degree in horticulture, you should take the appropriate college preparatory classes. Courses will include biology, chemistry, mathematics, physics, English, and foreign language. Classes in social science and humanities are required in most high school college preparatory programs. Computer courses and any classes that will help you to develop good communication skills should be considered also.

Extracurricular activities such as science clubs and science fairs will prove valuable even if you do not plan to attend college. Summer employment in any related area also helps, and might include working for summer camps, parks, gardens, greenhouses, florist shops, plant nurseries, and landscape design and maintenance contractors. Hobbies such as gardening, photography, art, camping, and computers are also useful.

College Curriculum

In today's employment market, a two-year college degree might be the minimum educational requirement for entry-level positions in some horticultural careers, but a higher level of education could be the minimum requirement for other fields. Technical positions in laboratories and in academic settings would require at least a four-year college degree for entry-level positions. Advancement or other positions could require a master's degree or Ph.D. There are also a few career options in horticulture for persons who do not have a college degree or technical school certificate. These positions are limited, however, and advancement could be more difficult and salaries more limited than for those persons with some education and training beyond high school. For more specific information on educational requirements, refer to the section covering your particular area of interest.

Technical schools and community colleges offer certificates and degrees in horticulture. The specific courses you take will depend on the curriculum of the school and your specific area of interest. Most of these schools will require courses in biology, math, chemistry, general horticulture, and English as part of a core curriculum. Additional horticulture courses can be selected depending on your specific interest. Most of these certificate and degree requirements can be completed within two years.

Universities and colleges offering degrees in most areas of horticulture are land-grant colleges and universities, which were established to teach the agricultural and technical sciences and are typically public colleges and universities. At least one such school is located in each of the 50 U.S. states.

In addition to core curriculum requirements such as English, mathematics, foreign language, chemistry, biology, social science, and humanities, your horticulture curriculum will include courses in general horticulture, plant physiology, entomology, plant pathology, and soil science. More specific courses are then determined by your particular career choice.

If you want to attend graduate school, you should take courses in writing, the arts, and social science. You should become proficient in using computers and a wide variety of software programs.

Summer employment, internships, and cooperative education provide invaluable work experience for students. Positions in government, college and university laboratories, agricultural and biological research stations, and private industry should be explored during the college experience.

If the curriculum permits, arrange to do an undergraduate research project under one of your professors or volunteer to assist one of your faculty members with his or her research. These kinds of experiences will not only teach you a lot about your chosen career field but can also help you to decide if a career in research and graduate school are the choices for you.

Other Personal Qualifications

Horticulturists need to be curious about the world around them and to be creative in solving problems. Most areas of horticulture require good writing and verbal communication skills, as well as physical strength and stamina. An interest in nature and the outdoors is beneficial. Other personal qualifications depend on the specific career path you choose. If your area of interest is in research and teaching, an understanding of the scientific method and attention to precision and detail are essential. Some areas require a flair for art and design, in addition to good interpersonal skills for working successfully with colleagues and customers.

Horticulturists are often required to understand and operate a wide variety of equipment, which could be as simple as a lawn mower or pruning shears, or as sophisticated as growth chambers or electron microscopes. These skills may be taught in courses in college, or they may be acquired by working in a laboratory or in field work.

Horticulture offers numerous interesting and rewarding career options, with varied work settings. The levels of education and the skills required to be successful in the field of horticulture are very broad, ranging from no college training to a Ph.D. and from the ability to operate a tractor to placing delicate flowers into an aesthetic arrangement. Because of this great diversity in horticulture, many people with different backgrounds, training, and interests can find a rewarding career in the many branches of the field.

SALARY AND BENEFITS

Salaries for horticulturists depend on the individual's level of education, the horticultural specialization, the geographic location of employment, and whether the person is employed in the private sector or by a government agency. An entry-level salary for a person without a college degree could be as low as the minimum wage, but in general average entry-level starting salaries range between $22,000 and $30,000 per year for a person with a bachelor's degree. As of this writing, the salary range for available horticultural positions with the federal government ran from $28,600 to $87,200.

Salaries for college teachers vary depending on rank, type of institution, and geographic area. According to a 2004–05 survey by the American Association of University Professors, salaries for full-time faculty averaged $68,505. By rank, the average was $91,548 for professors, $65,113 for associate professors, $54,571 for assistant professors, $39,899 for instructors, and $45,647 for lecturers. Overall, salaries tend to be lower in more rural or economically weak areas of the country, and as in other career fields, the higher your education level and the greater your skills, the better your salary and job opportunities.

There are more benefits to a career in horticulture than salary, however. Many jobs in horticulture provide rewards such as the freedom to work independently or with others who share your interests, a pleasant work environment indoors and outdoors, the opportunity to be creative and solve problems, and the possibility of travel. In addition, the great satisfaction and sense of accomplishment that come from working in a career field that strives to solve some of the problems associated with world hunger and environmental concerns, as well as the opportunity to create beautiful surroundings where people live and work, are probably some of the most positive rewards for pursuing a career in horticulture.

POMOLOGIST

The study of the production of fruit is called pomology, and those who work in this career field are called pomologists. Pomology is the oldest branch of horticulture in the United States. The first fruit tree nursery in America was started in 1730 in Flushing, New York by the Prince family, and the commercial production of fruit began in the mid-1880s when shipping long distances became feasible. It should be mentioned that the term pomology is a misnomer. Technically, *pome* is a term used to refer to specific fruits such as apples and pears, but the term pomology has come to be used to refer to the study and production of any fruit.

Fruit crops include deciduous tree fruits such as apples, pears, peaches, cherries, plums, and apricots; evergreen tree fruits such as lemons, limes, oranges, grapefruits, bananas, and pineapples; nut tree fruits such as pecans and walnuts; and small fruits such as blueberries, blackberries, raspberries, strawberries, cranberries, and table grapes (as opposed to wine grapes). The growing and processing of grapes for wine is called enology and is a highly specialized type of grape production.

Fruit crops are grown throughout the United States and the rest of the world with production concentrated in areas best adapted in soils and climate to the specific crop. For example, in America citrus production is concentrated in Florida, southern California, and parts of Texas, while the major areas for apple and pear production are located in the Pacific Northwest and around the Great Lakes. This type of specialization is found throughout the world.

Specific Work Performed

Pomologists propagate, plant, grow and maintain, and harvest and ship specific tree and small fruit crops. They prepare the planting areas based on the specific needs of the crop, establish the plants either by hand or with mechanical planters, water, fertilize, prune and train, harvest, package, and ship the fruit produced. Production pomologists monitor their crops for insects and diseases and apply appropriate pesticides.

Most producers specialize in a crop, but some of the larger fruit production businesses are highly diversified and produce many different crops. In many smaller fruit production operations, most of the tasks described are done by hand, but larger operations are highly mechanized and may require the operation of sophisticated equipment.

The intensity of the production process depends on the requirements of the crop and on whether it is produced for fresh consumption or for processing (canned, frozen, pickled, dried, and so on). For example, apples grown for applesauce do not need to be as blemish free, as perfectly shaped, as large, as colorful, or even as freshly picked as apples bought by the public in grocery stores.

Supervisory and management personnel in production pomology train and oversee field personnel, and are responsible for having all work performed correctly and safely. They may coordinate all phases of the operation, set budgets, purchase equipment, and arrange for support services such as harvesting, packaging, and shipping. Some may even be involved in the sales of the crop. Small businesses may require a single employee to perform several of these duties, but larger businesses typically have separate managers for the different divisions in production, sales, packing, storage, shipping, and general management. Some possible job titles you might find in production pomology include grower, field supervisor, operations manager, field technician, and shipping supervisor, among others.

Pomologists also may be employed in research positions by private industry, government, and colleges and universities, while some occupy teaching positions. Research pomologists study one specific fruit, or a category of fruit crops, to determine methods of breeding, propagating, and growing higher-quality plants more efficiently and for a greater, and thus more profitable, yield. They test these methods both in the laboratory and in field production. Many work closely with commercial growers, addressing concerns or problems specific to that grower.

The nature of their work requires research pomologists to operate many different kinds of laboratory and field equipment. These might include sophisticated tissue culture and analysis equipment as well as more mundane equipment such as tractors, mechanical harvesters, or even hand pruners.

Pomologists involved in teaching at the college and university level often conduct research as well, and many consult with commercial growers on problems of mutual concern.

Specific Skills Required

People employed in the field of pomology must be knowledgeable about, and skilled in, the propagation, growing, harvesting, and shipping of fruit crops. Most pomologists specialize in a particular fruit crop or area of fruit production such as small fruits or deciduous tree fruits, and must have a thorough knowledge of the cultural requirements, harvesting, shipping, and handling through postproduction of that specialty crop.

Pomologists must be able to operate equipment associated with the production of their specific crop, such as row cultivators, mechanical harvesters, pesticide sprayers, and other such equipment. They're required to have a thorough knowledge of soils, irrigation practices, fertilization procedures, harvesting techniques, and pruning and training techniques as well. Knowledge of pest identification and control is also essential in this career field.

Since anyone in a management or supervisory position in production pomology must be able to teach others how to perform the specific tasks associated with the business, good leadership and oral communication skills are important in such jobs. A speaking knowledge of Spanish is helpful because most laborers employed in the commercial production of fruits are Mexican or Central American. General business skills are also key.

In addition to the specific skills required of a commercial pomologist, those employed in research or academic settings must be familiar with a variety of laboratory and field equipment and the numerous techniques used. A more in-depth knowledge of plant physiology, biochemistry, pathology, and genetics is needed for success in the research branch of pomology. A good understanding of scientific methodology, statistics, and excellent oral and written communication skills are also required.

Work Settings

Pomologists are employed in a variety of work environments, such as orchards, typical farm settings, and even tropical plantation operations. Some commercial operations are small, family-owned businesses, while some are large multinational conglomerate corporations. Commercial operations might employ only a few workers or hundreds of them. Today, there is a trend among growers toward forming cooperative associations in order to increase capitalization, storage capabilities, and marketing and shipping possibilities.

Commercial fruit production employees work outdoors, sometimes in inclement weather. Hours are irregular, and stressful seven-day weeks, full of 12-hour days, are common during peak seasons. Often, work is not seasonal as in many other horticulture fields because many tasks required in fruit production, such as pruning and some pest control practices, are best performed in winter months. Because fruit production is related to the climatic requirements of the specific crop, your particular crop area may be the only factor affecting your geographic location. For instance, if you are interested in growing pineapples or bananas, your work will be performed in tropical locations. If apples are your primary interest, you will work in temperate and even cold geographic areas.

Commercial fruit production operations are located near large urban centers or in very rural settings, depending on the crop produced. Climate permitting, crops with short shelf lives or poor handling or shipping characteristics are grown close to the point of sale to reduce loss and production costs. Crops grown for processing are produced near the processing facility, while crops that store and ship easily can be produced in a wider variety of areas, again climate permitting. Thus, employees in this field have many choices for the location of their work.

Research and teaching pomologists work in traditional research and academic settings. However, as in many other horticulture career fields, these scientists also work outdoors.

Pomology production and research usually require physical labor as a routine part of the job. This labor can vary from highly strenuous to relatively nonstrenuous. Bending, stooping, lifting, and standing for long periods are common, and accommodations can be made for persons with physical handicaps.

Employment Outlook

Food production is the largest and fastest growing area of horticulture. Increased public concern for personal health and for the environment has caused the demand for healthy nutritional food crops to increase. Fruit consumption has become more popular today because of the low-fat, low-calorie characteristics of many fruits. The food pyramid released by the Departments of Agriculture and Health and Human Services emphasizes the importance of incorporating fruits into our daily diets. In addition, interest from the grower as well as the consumer in production techniques to increase yields and profitability will continue to increase the need for skilled growers and highly qualified researchers and academics to continue looking for solutions to production and other public concerns. Thus, the need for highly skilled and well-trained pomologists will remain high.

Advancement Opportunities

Advancement in pomology depends on your level of education and skills, whether you are employed in a commercial or research/academic setting, and the size of the operation. In commercial pomology, skills are very valuable; thus, many people begin their careers as laborers or lower-level management or supervisory personnel. As their education level increases and their skills improve, workers advance into more responsible positions such as grower, field supervisor, and general manager. Persons with a college degree may enter the profession as supervisors, graders, sales managers, and even assistant or head growers. Typically, employees in this field progress from assistant-level positions to more and more responsible management positions. Some pomologists may even own their own fruit production operations.

Advancement in research and academic settings depends on your education level as well as your skills. Those with a minimum of a bachelor's degree typically enter the field as laboratory assistants or technicians, and progress as their education level increases or their skills improve, possibly to managing a laboratory or research project with other technicians under their supervision. Further advancement would require additional education or a higher degree. Pomologists in teaching positions follow the typical career path from assistant professor to associate professor to full professor, depending on degrees held and their personal abilities and ambition.

Education and Training Requirements

Competition for positions in pomology production is increasing as the field becomes more and more technical and as the size of fruit production operations continues to change from family-owned farms/orchards to the larger, diversified, and highly capitalized international corporations. Today, a minimum of an associate's degree in pomology is recommended for students who wish to advance in this field to career-level positions. Obviously, a bachelor's degree will make you even more competitive for entry-level positions, and advancement should occur more quickly, depending on your skills. This is especially true if you are interested in employment with large commercial fruit production operations.

High school students interested in pursuing this career field should take college preparatory courses in biology, chemistry, mathematics, writing and oral communication, business management, and computer skills. Working part-time in any agricultural setting provides exposure to techniques and equipment that will prove to be helpful.

At the community college and college and university levels, students will be required to take a core curriculum, plus courses in general horticulture, soils, plant pest identification and control, plant nutrition, writing and oral communication, as well as specialized courses in fruit production. To obtain a master's or doctorate degree, students will complete course work in a more specific area of pomology, such as pome fruit or tropical fruit production, and complete a research project and major paper under the supervision of a faculty advisor and faculty committee. Additional course work will be required in areas such as statistics.

Internships, summer employment, and cooperative education experience are helpful in gaining practical experience and training in pomology. Many opportunities will be found in small fruit production operations, commercial settings, colleges and universities, and even fruit processing plants. Although certain degrees and knowledge are required for some levels of employment in this field, like many of the other production-oriented branches of horticulture, success in pomology can depend heavily on your skills and abilities in the field rather than on any specific education level or training.

Other Personal Qualifications

A love of nature and the outdoors is essential for anyone interested in this field. In addition, pomologists need to have a concern for detail and accuracy in their work. Also, they should be observant, patient, and able to work with a wide variety of people from diverse backgrounds.

Salary

As in many horticulture occupations, specific salary information for pomologists is limited and rarely reported by reliable sources. Data are based on general information about agriculture workers' salaries and is reported without reference to factors such as size of business, geographic location, and experience and education levels of employees.

Entry-level laborers should expect to earn little more than minimum wage. However, those with experience and/or an associate's degree in pomology can expect salaries in the low to mid-20s. As they advance to supervisory and management positions (such as field managers and growers), their salaries will increase to around $35,000 annually. A master's degree can increase your income up to $48,000 per year in private industry and slightly higher in government employment. Both faculty teaching and pomology research are paid the prevailing salaries of higher education institutions, as described earlier.

Other Rewards

Pomology careers might also offer the opportunity for travel. Working outdoors, with nature, and with a variety of people can be rewarding as well. Knowing that your work is important in improving the lives and well-being of other human beings throughout the world is a primary reward in pursuing a pomology career.

FLORICULTURIST, GREENHOUSE GROWER, GREENHOUSE MANAGER

Floriculture has been an important part of the horticulture industry for a long time and accounts for approximately one-half of the nonfood horticulture industry in the United States. It is primarily a greenhouse industry except in geographic areas where the climate is mild and conducive to the growing of specific floriculture crops. The field is highly competitive with very narrow profit margins, and the industry has developed into a group of highly specialized businesses. Greenhouse floriculture is probably the most highly technical and sophisticated kind of horticulture.

The study of growing and marketing flowers and foliage plants is called floriculture, and persons working in this field are called floriculturists. Floriculturists are involved in the growing of cut flowers such as roses, snapdragons, orchids, and carnations; flowering and foliage potted plants such as chrysanthemums, poinsettias, lilies, tropical palms, fig trees, and philodendrons; and bedding plants such as petunias, marigolds, impatiens, and geraniums. Job titles associated with this career field include propagator, grower, production manager, marketing manager, salesperson, technician, and inventory controller.

Specific Work Performed

Floriculturists are involved in the propagation and growing of cut flowers, flowering potted plants, foliage plants, and bedding plants. They plant, prune, water, fertilize, harvest, and ship these plants, as well as monitor and manipulate the environment under which they are grown both indoors in greenhouses and outdoors in field production where the climate is favorable. Floriculturists also conduct research to improve the production and handling of floral and foliage crops for greater yields and improved quality. Some may teach at the college and university level, and some, such as commercial floriculturists, may market their crops to retailers or to the public directly. Others are directly involved in the daily activities of producing their crops, while still more may hold supervisory and management positions. Many, meanwhile, are responsible for the scheduling of crop production and for the mechanical operation of the greenhouse.

Growers are generally responsible for all stages of production of a single crop, a portion of a crop, or a group of crops. Large floriculture operations may employ several growers, who must train and supervise workers. Production managers are responsible for supervising growers, and they coordinate all production activities with efficiency and profitability in mind. Marketing managers, on the other hand, oversee the grading, handling, storage, packing, and shipment of crops. They manage warehouse personnel who handle plants from harvest, through storage in coolers, to packaging and shipping. They also supervise sales personnel and coordinate shipping. Inventory controllers, meanwhile, are responsible for the coordination of production between marketing and production departments and typically buy all supplies and equipment.

After harvesting, graders sort flower crops into groups based on quality standards set by the industry. Propagators are responsible for increasing the number of plants in production by growing plants from seed and employing other techniques such as cutting and division, while technicians are most often employed in research and academic settings and assist or actively participate in the research of faculty members.

Specific Skills Required

People employed in the field of floriculture must be knowledgeable about, and skilled in, the propagation of plants as well as the growing and maintenance of those plants until they are shipped to buyers. Floriculturists must be familiar with a wide variety of crops, their specific cultural requirements, and their timing, harvesting, shipping, and handling through postproduction. They should be able to operate equipment commonly found in greenhouses, such as pruners, sprayers, pH meters, soil sterilizers, and fertilizer injectors, in addition to heating, cooling and ventilating, watering, and humidifying equipment. They should be familiar with using a variety of light, temperature, and humidity monitoring devices, from something as basic as a thermometer to more highly sophisticated and complex equipment such as growth chambers. Many floriculture operations today are partially computerized; thus, the ability to operate a computer and use basic software is helpful.

Floriculturists must be knowledgeable about soils and the unique soil mixes used in the production of floriculture crops. They must be familiar with standard potting and transplanting techniques and the specialized equipment designed to fill containers with soil mixes for high volume production. They should be able to identify typical floriculture crop pests and diseases, and be acquainted with proper control techniques and the operation of pest control equipment. Anyone in a management or supervisory position in the floriculture production section of the field must be skilled in all of the tasks required for production and be able to teach others how to perform these tasks as well.

General business and management skills are very important in floriculture production, and a thorough understanding of greenhouse construction and the unique aspects of the operation of floriculture production environments is critical, especially for those in management and supervisory positions. A firm grasp of the specifics of heating, cooling, and controlling other climatic factors such as humidity, and even light, is essential for the floriculturist. Today, the growth of many floriculture crops is regulated by chemical and mechanical methods, and the floriculturist must be adept at the application of these chemicals and the recommended mechanical techniques.

In addition to the specific skills required of a commercial production floriculturist, those employed in research or in academic settings must be familiar with a variety of laboratory techniques and equipment, including microscopes (such as those ranging from simple binocular equipment to electron scanning microscopes), cell and tissue culture, staining techniques, and a wide variety of other laboratory techniques. Floriculturists employed in these areas probably should have a more in-depth knowledge of plant physiology and biochemistry than those employed in production. An understanding of standard statistical methods, scientific methodology, and good written and oral communication skills are also required.

Work Settings

Floriculturists are employed in a variety of work settings, including production greenhouses and outdoor field production areas in places where the climate is mild or conducive to the production of certain crops. Commercial greenhouses and field production are generally limited geographically to areas where the climate provides the best cultural environment, perhaps offering high light conditions and warm or moderate temperatures. Typically, these operations are also located near but outside of urban areas where a qualified labor pool is available, transportation costs are low, and land is available for agricultural use. States such as California, Florida, and Texas account for the majority of commercial flower production in the United States, but commercial production is found at some scale in all 50 states. Currently, the majority of cut flowers produced in the world come from Holland and Central and South America. If you are interested in traveling and working overseas, this aspect of floriculture production or research may be an ideal choice for you.

Commercial floriculture operations in the United States began as family-owned businesses and many remain so today. However, their size can range from one or two acres of production to several thousand times that amount. In fact, today many commercial floriculture businesses are subsidiaries of large conglomerate corporations.

Commercial production includes growing flowers and other crops for both wholesale and retail sales. In commercial production settings, employees may be required to work in adverse weather conditions such as extreme heat or rain, or they may work in greenhouses where temperatures can rise and fall dramatically and humidity levels are extreme. Workers involved in the postharvest handling occasionally work in coolers where temperatures are low and humidity is very high.

Floriculturists are also employed in research greenhouses and laboratories by private industries, academic institutions, and a variety of government agencies. Work environments can range from typical outdoor settings and greenhouses encompassing hundreds of acres similar to those in commercial floriculture production to confined growth chambers and sterile laboratories. Generally, floriculturists work with other people, but researchers may work alone or in contact with a limited number of technicians and other researchers. Physical demands for researchers may be similar to those required of production floriculturists; however, work environments in laboratories generally are less physically demanding and more "officelike."

Work hours in commercial production and floriculture research are often long and irregular. The demands of the crops and the research determine the work schedule, and frequently timing is critical for various aspects of production and for collecting research data. Also, commercial floriculturists are often employed in settings involved in the production of holiday crops such as poinsettias and lilies, and they must work long and stressful hours for certain peak production times such as Christmas, Valentine's Day, Easter, and the spring bedding plant season.

Employment Outlook

The employment outlook for floriculturists is good. Other than areas involved in the production of food, floriculture is the largest employer in the field of horticulture. Public concern for the environment and an overall improved quality of life has caused the demand for floriculture crops to increase steadily. As this demand continues to increase, so will the need for skilled floriculturists in commercial production settings in order to obtain high-quality and marketable products. As production demands increase, so will the need for qualified researchers and academics to look for methods to increase production yields, improve flower quality and shelf life, reduce production cost, improve profitability, and train future floriculturists.

Advancement Opportunities

Advancement in floriculture will depend on whether you are employed in a commercial or academic setting. In commercial floriculture production, experience is highly desirable; thus, many people begin their careers as greenhouse or field production laborers in order to learn as many aspects of the industry as possible. They progress into positions such as assistant growers, production managers, and general managers as their skills improve or their education level rises. Those with a college degree in floriculture may bypass the laborer stage and enter the profession as assistant growers, graders, or marketing personnel and advance into supervisory and management positions as well. Some experienced floriculturists may even own their own greenhouses or field production businesses.

Advancement in research and academic settings depends primarily on your education level. Those with a minimum of a bachelor's degree in floriculture may enter the field as a laboratory assistant and progress to assistant technician, then technician. Further advancement requires a higher level of education. Floriculturists in college teaching follow the typical career path from instructor to assistant professor to associate professor to full professor, depending on personal ability and degrees held.

Education and Training Requirements

Unless you think you will be satisfied with a laborer's job, competition for positions in floriculture will require that you have a minimum of an associate's degree for an entry-level position in this career field. A bachelor's and master's degree may be required in some industrial settings, and if research and/or teaching interest you, a master's or Ph.D. is the minimum requirement.

Students interested in pursuing a career in floriculture should take the required college preparatory courses in high school, as well as classes in biology, chemistry, mathematics, writing and oral communication, business management, and computer skills. Part-time and summer work in wholesale or retail greenhouses, nurseries and garden centers, florist companies, or floriculture research laboratories are advisable to gain as much experience as possible in the field.

At the community college, college, and university levels, students are required to take the core horticulture curriculum with additional or elective courses in greenhouse management and crop production, herbaceous plant material identification and production, pest identification and management, business management, and writing. Students pursuing a master's and/or doctoral degree should concentrate additional course work in a specific area of floriculture—such as cut flower production, bedding plants, or foliage plants—and must complete a research project and a thesis paper or dissertation in the specific area of study.

Internships, summer employment, and cooperative education work experience are helpful to gain practical experience in floriculture. These types of work experience can also help you decide which of the many subdivisions of floriculture interests you most.

Other Personal Qualifications

People interested in floriculture as a career field should be detail oriented and accurate in their work. They must be patient, analytical, and observant. Creativity and artistic flair is also helpful. As in any horticulture career field, a love of nature is essential. Most floriculture production involves physical labor—at times strenuous—requiring bending, stooping, lifting, and standing for long periods. As with other occupations, accommodations for persons with physical handicaps can be made.

Salary

Salaries in floriculture are as variable as the tasks performed and skills required. Specific data are sketchy and rarely reported by reliable sources. Typically, salaries in this career field can begin at minimum wage for uneducated and unskilled laborers, and some small production greenhouses pay little more for any entry-level position. The more education, experience, and skills you have, the higher the salary you can expect to be offered. As responsibilities, education, and experience increase on the job, salaries will increase also.

Annual starting salaries for those with bachelor's degrees range from $30,000 to $40,000. The starting salary range for floriculturists with master's degrees is from $35,000 to $50,000, while Ph.D. holders start in the $60,000 to $70,000 salary range.

Salaries will vary geographically because of the availability of labor and the prevailing wage for a given area. Also, as in many other fields of horticulture, salaries may be based on skill and experience rather than set criteria such as education level and job title. Remember, as stated earlier, these data were inconclusive and the student would be wise to canvas employers in his or her geographic area, and in his or her specific area of interest, to determine the expected salary level.

Other Rewards

Anyone who enjoys working with nature and in an outdoor or simulated outdoor environment will find the career field of floriculture rewarding. Much satisfaction can be derived from producing an aesthetic and marketable product from start to finish, either alone or with the cooperation of a company team. The mental challenges of working with a living product can be stimulating, and knowing that your efforts bring pleasure to others is highly rewarding. To those interested in travel, a floriculture career can also offer benefits such as trips to Central and South America.

OLERICULTURIST

The study of the production of vegetables is called olericulture, and those who work in this career field are called olericulturists. Most vegetables are grown as annuals, and shifts in crops produced occur frequently and can be easily made. In the past, a large part of the vegetable production industry was diversified, but current trends are toward specialization due to rising production costs.

Vegetable crops include tomatoes, lettuce, potatoes, carrots, sweet peas, corn, onions, and squash, to name but a few examples. Note that there is no clear-cut definition for a vegetable and no clear delineation between a fruit and a vegetable. In botany, a tomato is technically a fruit, but in production a tomato is considered a vegetable. For our purposes, we will define a vegetable as any edible portion of an herbaceous plant consumed principally during the main portion of a meal. Some vegetables are valued for their botanical fruits—tomatoes and squash are examples—while other vegetable crops are valued for their edible leaves, stems, or roots. Examples of the latter include spinach, potatoes, and carrots. Most vegetables have their origins in the Old World, but crops such as potatoes, corn, peppers, and tomatoes were discovered in the Americas and exported to Europe where they became diet staples.

Vegetable crops are grown throughout the United States and the rest of the world. Large-scale production in the United States is concentrated in specific states such as California, Texas, and Florida because of numerous cultural and business factors, including but not limited to labor availability, land costs, transportation, climate, and the length of the growing season. However, vegetable crops are produced to some extent in all states, indicating the flexibility of the industry.

Vegetables are grown for fresh consumption or for processing, and most of those crops produced for long-distance or world trade are processed with an eye to their perishability and bulk. Fewer vegetable crops than fruit crops enter the world market, but many are grown for local sale or short-distance transport and do not require processing prior to sale.

The vegetable industry in the United States may be classified into the categories of market gardening and truck gardening. Market gardening refers to the production of a variety of vegetables for local and roadside markets, generally near large population centers. This type of growing is gradually disappearing in the United States because of land costs and improved shipping techniques. Truck gardening, on the other hand, refers to the large-scale production of a few crops for wholesale markets and for shipping. It is dependent on climate, soils, and suitable growing seasons rather than market proximity. In many other countries, another classification of vegetable production is home gardening, particularly in developing countries where the home garden remains a primary source of food.

Specific Work Performed

Olericulturists propagate, plant, grow and maintain, and harvest and ship a variety of vegetable crops. They prepare the planting areas based on the specific needs of the crop, establish the plants either by hand or with mechanical planters, water, fertilize, harvest, package, and ship the vegetables produced. Producers monitor their crops for insects and diseases and apply appropriate pesticides.

Large-scale producers tend to specialize in one crop, particularly in geographic areas where the climate is the same throughout most of the year (southern California, for example), but many smaller vegetable producers are diversified, typically rotating two to three crops per year. For example, in the mid-Atlantic states, cold crops such as broccoli and/or cabbage are rotated with a warm crop such as tomatoes or corn. In many smaller vegetable production operations, most of the tasks described are done by hand, but larger operations are highly mechanized and may require the operation of sophisticated equipment.

The intensity of the production process depends on the requirements of the crop and on whether it is being produced for fresh consumption or for processing (canned, frozen, pickled, dried, and so on). For example, tomatoes grown for canning do not need to be as blemish free, as perfectly shaped, as large, as colorful, or even as freshly picked as tomatoes bought by the public in grocery stores.

Supervisory and management personnel train and oversee field workers and are responsible for ensuring all work is performed correctly and safely. They may coordinate all phases of the operation, set budgets, purchase equipment, and arrange for support services such as harvesting, packaging, and shipping. Some may even be involved in the sales of the crop. Small businesses might require a single employee to perform several of these duties, but larger businesses typically have separate managers for the different divisions in production, sales, packing, storage, shipping, and general management. Some possible job titles you might find in production olericulture are grower, field supervisor, operations manager, field technician, storage supervisor, packing and grading supervisor, shipping supervisor, salesperson, broker, and others.

Olericulturists may be employed in research positions by private industry, government, and colleges and universities. Some find jobs in teaching as well. Researchers typically study one specific category of vegetables to determine methods of breeding, propagating, and growing higher-quality plants more efficiently and for a greater, and thus more profitable, yield. They test these methods both in the laboratory and in field production. Many work closely with commercial growers, addressing concerns or problems specific to that grower.

The nature of their occupation requires research olericulturists to work with many different kinds of laboratory and field equipment. These might include sophisticated tissue culture and analysis equipment as well as field equipment such as tractors, mechanical planters and harvesters, boom sprayers, or even hand pruners and trowels.

Olericulturists involved in teaching at the college and university level often conduct research, also, and may consult with commercial growers and government agencies on problems of mutual concern.

Specific Skills Required

People employed in the field of olericulture must be knowledgeable about, and skilled in, the propagation, growing, harvesting, and shipping of several vegetable crops. They must have a thorough knowledge of cultural requirements, harvesting, shipping, and handling through post-production of the crop produced.

Vegetable producers and researchers must be able to operate a wide variety of equipment associated with the production of their specific crop, such as row cultivators, mechanical planters and harvesters, pesticide sprayers, irrigation and fertilization equipment, and other such equipment. They must have a thorough knowledge of soils, irrigation practices, fertilization procedures, harvesting techniques, and pest identification and control techniques. Like most professionals in the horticulture industry, vegetable producers and researchers work with hazardous chemicals and must be knowledgeable in the safe application of these chemicals.

A good foundation in business management is beneficial to those employed in the production area of olericulture. Anyone in a management or supervisory position in production olericulture must be able to train others to perform the specific tasks associated with the business. Thus, good leadership and oral communication skills are important. A speaking knowledge of Spanish is also beneficial because most laborers employed in this industry are Mexican or Central American.

In addition to the specific skills required in commercial vegetable production, those employed in research or academic settings must be familiar with a variety of laboratory and field equipment, as well as the various techniques associated with them. A more in-depth knowledge of plant physiology, biochemistry, pathology, and genetics is needed for success in research. Understanding scientific methodology, a background in statistics, and excellent oral and written communication skills are required to record, interpret, and report research data.

Work Settings

Olericulturists are employed in a variety of work settings such as greenhouses, typical farm settings, commercial and university laboratories, and many government agencies in the United States and overseas. Some commercial operations are small, family-owned businesses, while others are large multinational conglomerate corporations. Commercial operations might employ only a few workers or thousands of them in various locations. Today, the trend is toward cooperative associations among small-scale producers to increase capitalization, as well as marketing and shipping capabilities.

Commercial vegetable production employees work outdoors, often in inclement weather, and in the interior environments of greenhouses. Researchers have long employed the latter for climate-controlled scientific study. The commercial greenhouse production of vegetables is a small but increasingly important part of the industry, particularly in the Midwest and Northeast.

Work schedules are long and irregular. Seven-day weeks and 12-hour days are common during peak planting and harvesting seasons. Often, work is seasonal, as in many other horticulture fields, especially in areas where the growing season is limited by weather. Because of the wide variety of vegetable crops produced throughout the world, your geographic location is limited only by your interests. Commercial vegetable production operations may be located near large urban centers or in very rural settings. Climate permitting, crops with short shelf lives or poor handling or shipping characteristics are grown close to the point of sale to reduce loss and production costs, while crops that store and ship easily can be produced in a wider variety of areas. Thus, employees in this field have many choices regarding their work location, both in the United States and overseas.

Research and teaching positions in olericulture can be found in traditional research and academic settings. However, as in many other horticulture career fields, these scientists also work outdoors in the field or greenhouse.

Vegetable production and research usually require physical labor as a routine part of the job. This labor is generally strenuous, and typically includes bending, stooping, lifting heavy objects and equipment, and working on your feet for long periods of time. Working in extreme heat, humidity, and even cold temperatures occurs frequently, depending on the time of year and on the requirements of the crop.

Employment Outlook

Food production is the largest and fastest-growing area of horticulture. Increased public concern for personal health and for the environment has caused the demand for healthy nutritional vegetable crops to increase. The current trend toward vegetarianism, as well as the heightened awareness of the nutritional and health values of vegetables, should result in an increased demand for new and improved vegetable crops and crop production methods in the coming years. In addition, interest from both the grower and consumer in production techniques in order to increase yields and profitability is likely to increase the need for skilled growers, and spur highly qualified researchers and academics to continue looking for solutions to production problems and ways to address public concerns. Thus, the need for highly skilled and well-trained vegetable growers and researchers should remain high.

Advancement Opportunities

Advancement in this career field depends on whether you are employed in a commercial or research/academic setting or a large or small operation, and depends on your level of education and skills. In commercial vegetable production, skill is very valuable; thus, many people begin their careers as laborers or lower-level management or supervisory personnel. As their education levels and skills improve, they advance into more responsible positions such as grower, field supervisor, and general manager, or into sales. Those with a college degree may enter the profession as supervisors, graders, sales managers, and even assistant or head growers. Typically, people in this field progress from assistant-level positions to more and more responsible management positions.

Advancement in research and academic settings depends on your education level as well as your skills. Those with a minimum of a bachelor's degree typically enter the field as laboratory assistants or technicians and progress as their education level increases or skills improve, possibly becoming managers of laboratories or research projects with other technicians under their supervision. Further advancement requires additional education and eventually a higher degree. Olericulturists in teaching positions follow the typical career path from assistant professor to associate professor to full professor, depending on the degrees held and their personal abilities and ambition.

Education and Training Requirements

Competition for positions in olericulture is increasing as the field becomes more and more technical and as the size of production operations continues to change from the family-owned farm operation to the large, diversified, and highly capitalized international corporation. Today, a minimum of an associate's degree in horticulture with a concentration in olericulture is recommended for students who wish to advance in this field to career-level positions. A bachelor's degree will make you more competitive for entry-level positions and advancement, depending on your personal skills. This is especially true if you're interested in employment with large commercial vegetable production operations.

High school students interested in pursuing olericulture as a career should take the appropriate college preparatory courses in biology, chemistry, mathematics, writing and oral communication, business management, and computer skills. Working part-time in any agricultural setting is also helpful because it exposes you to field techniques and equipment.

Students at the community college, college, and university levels are required to take a core curriculum plus courses in general horticulture, soils, plant pest identification and control, plant nutrition, writing, and oral communication, as well as specialized courses in vegetable production. To obtain a master's or doctorate degree, students must finish course work in a specific area of olericulture, such as tomato production, and complete a research project and major paper under the supervision of a faculty advisor and faculty committee. Additional course work is required in areas such as vegetable physiology, genetics and breeding, biochemistry, and statistics.

Internships, summer employment, and cooperative education experience are helpful in gaining practical experience and training in this career field. Many opportunities exist in small vegetable production operations, commercial settings, colleges and universities, and even food processing plants. Although certain degrees and knowledge are required for some levels of employment in this field, like many of the other production-oriented branches of horticulture, success in olericulture can depend heavily on skills and abilities in the field rather than on any specific education level or training.

Other Personal Qualifications

A love of nature and the outdoors is essential for anyone interested in this career. Field and research olericulturists should also recognize the need for detail and accuracy in their work. In addition, they should be observant, patient, and able to work with a wide variety of people from diverse backgrounds.

Salary

As in many horticulture occupations such as floriculture and pomology, specific salary information is limited and rarely reported by reliable sources. Data are based on general information about agriculture workers' salaries and are reported only as averages. Again, this information is misleading because factors such as size of business, geographic location, and the experience and education levels of employees are not reported with the salary data.

Entry-level laborers should expect to make little more than minimum wage. However, those with experience and/or an associate's degree in horticulture can expect average salaries in the low to mid-20s. As they advance to supervisory and management positions (such as field managers and graders), their salaries should increase to about $30,000 to $40,000 annually. A master's degree can increase income up to $48,000 per year in private industry, and slightly higher than that in government employment. Teaching and research faculty, meanwhile, are paid the prevailing salaries of higher education institutions. According to a 2004–05 survey by the American Association of University Professors, salaries for full-time faculty averaged $68,505. By rank, the average was $91,548 for professors,

$65,113 for associate professors, $54,571 for assistant professors, $39,899 for instructors, and $45,647 for lecturers. Faculty salaries vary widely according to the type of institution, its location, and its reputation.

Other Rewards

Olericulture careers may also offer the opportunity for travel, particularly for those employed in research and academic positions. Such travel may not be especially exotic, however, given that the demand for consultants in this area is highest in third-world countries. Working outdoors, working with nature, and working with a variety of people, typical in this field, can be rewarding also. Knowing that your work is important in improving the lives and well-being of other human beings throughout the world is perhaps the primary reward of pursuing a career in olericulture, however.

NURSERY GROWER, MANAGER

In the United States, the production of fruit crops such as apples, peaches, or even citrus fruits led to the development of the nursery industry as we know it today. Although many nurseries do produce fruit trees, the majority of nursery businesses today grow perennial ornamental plants such as shade and flowering trees, evergreen trees, deciduous and evergreen shrubs, and ground covers. A person who produces or distributes these ornamental plants is called a nursery grower or nursery manager, and a person who conducts research or teaches in this area is referred to as an ornamental horticulturist.

The nursery industry can be divided into two categories: wholesale production nurseries and retail nurseries or garden centers. Job titles in this career field depend on the type of nursery operation, however. Production nurseries typically employ plant propagators, inventory controllers, field laborers, field foremen, field supervisors, managers, salespeople, sales managers, and shipping and receiving foremen. They also occasionally employ brokers, who are independent, self-employed sales agents who locate and purchase plants from other production nurseries to fill customer orders. Retail nurseries or garden centers employ salespeople, managers, and buyers. Occasionally, these nurseries offer other services such as design and installation, thus creating opportunities for designers and the appropriate construction personnel.

Specific Work Performed

Plant nurseries propagate, grow, and sell trees, shrubs, and other plants. The production of young fruit trees, some perennial vegetables (asparagus, for example), some flowers, herbs, and small fruit plants, as well as evergreens for Christmas trees are also an important part of the nursery industry. Nurseries may sell this plant material wholesale or retail directly to landscape contractors, to garden centers, and to landscape architects or designers. Wholesale nurseries usually limit their production to a relatively few types of plants to maximize efficiency, but some large production nurseries may grow several hundred species of plants.

Some nurseries sell retail directly to the consumer. These retail nurseries often provide other related services such as the design, planting, or maintenance of the plants and have a garden center as a part of their operation where they sell fertilizers, pesticides, tools, and other related materials and products. These nurseries may grow a small portion of their own plants and buy additional stock from wholesale nurseries. Some nurseries, however, limit their sales strictly to a mail order/catalog business. Nursery growers/managers are in charge of the overall operation of these facilities or may supervise a specific portion of it.

Work as a nursery grower/manager includes a wide variety of tasks: plant propagation through seed, grafting, cuttings, and even cell or tissue culture; greenhouse operation and management; the preparation and maintenance of outdoor growing fields or container production areas; weed, disease, and insect control; plant breeding; the general monitoring, pruning, and growing of the plants; digging; loading and transporting trees, shrubs, and other plants; and sales and other business operations. In larger nurseries, these tasks are typically performed by a number of personnel, ranging from the supervisory to the general laborer level. In smaller nurseries and in many owner-run operations, the nursery manager performs all of the functions with the help of a few laborers.

Working in the nursery industry may involve seasonal employment depending on the level of your position and the geographic location of the nursery. In areas where the climate is generally warm or mild, year-round employment is normal, even necessary, for a profitable operation. In areas where winter weather may force outdoor work to stop or slow down, many laborers and some field personnel will be laid off for a brief period. Management, sales, and supervisory employees are not generally affected by the seasonal nature of outdoor nursery work.

Specific Skills Required

Thorough knowledge of plant materials and their requirements for proper growth and development into a salable product is essential. An understanding of soils, irrigation practices, pruning techniques, and pest control practices is important as well. Business management skills and sales ability are also valuable. Good interpersonal skills are essential in dealing with the wide variety of employees and customers a nursery manager or grower will encounter. Physical stamina and the ability to operate heavy machinery and equipment, as well as lift heavy loads, may also be required.

Work Settings

Nursery managers and growers are employed in a wide variety of settings. Some work in sterile laboratories propagating plants by cell and tissue culture, while some acquire jobs in greenhouses propagating and growing containerized and other plant materials. Many nursery managers and growers work exclusively outdoors in the nursery under a wide variety of weather and physical conditions. They work alongside a wide range of employees with diverse backgrounds, from uneducated and inexperienced but hardworking field personnel to highly educated and skilled technicians, managers, and field employees.

Employment Outlook

Nursery work is a growing field of employment at all levels. Openings exist in small local nurseries, large wholesale and retail operations, and in research departments at small and large corporations. Government employment is also available at the state and national levels. State land-grant universities and community colleges employ people with a background in nursery operations and management in positions such as field and laboratory technicians and nursery managers.

Advancement Opportunities

Advancement in this career field depends largely on ambition, experience, and the willingness to work long hours during certain times of the year. Many people start in this field as laborers and work their way up to such positions as field foremen, supervisors, managers of particular operations or crops, salespeople, office workers, and other titles. Supervisory and management positions are available to those who acquire a broad range of knowledge and experience. Many nursery managers, in fact, advance their careers by opening their own businesses.

Education and Training Requirements

The level of education and training required for a successful career as a nursery manager or grower depends on the specific level of employment and the specific position held. Most employers prefer some formal education and training. High school courses in science, mathematics, mechanics, art, and business are a good way to prepare for this field. Foreign language mastery is also helpful, especially Spanish. Basic computer skills is also expected.

Even for entry-level positions, education and training beyond high school is advantageous. Two-year degrees in nursery management, nursery crop production, and general horticulture are available from junior and community colleges and technical and vocational schools. These degrees and certificates provide access to entry-level management/supervisory trainee positions in small- to medium-sized nursery businesses. Larger companies, on the other hand, may require a four-year degree in nursery management, nursery crop production, or general horticulture. At either education level, course work should include biology, chemistry, mathematics, writing and communication skills, and computer operations, along with the other general core course requirements of the institution. The specific horticulture classes taken will depend on the particular requirements of the college or university.

Research and teaching in this career field probably requires a minimum of a master's degree or even a doctorate. Specific course work varies according to the exact nature of the degree obtained and the institution granting it. Some people in this field hold related diplomas in areas such as chemistry, agronomy, entomology, soil science, and biology.

Because training and experience are as important as education in nursery management, students are encouraged to seek summer employment, internships, or cooperative education work experience with nurseries, landscape contractors, and nursery industry–related laboratories. (See Chapter 4 for information about landscape contractors.)

Other Personal Qualifications

Curiosity about and a love of plants is a basic requirement for anyone in this career field. Good health, average strength, manual dexterity, and a good sense of design are essential. Business management skills, sales ability, patience, and the ability to interact successfully with the public are critical in the field of nursery management.

Salary

Earnings in this field vary widely depending on the size and location of the business, as well as your education and training. Some employees are paid an hourly wage ranging from the federal minimum wage to $15.00 or more. Average salaries for nursery managers vary from $32,000 to $40,000 per year, but can be higher depending on the size of the operation. Highly educated and trained personnel in teaching or research can earn $60,000 or more per year depending on the institution. Those employed in sales often work on a commission basis, especially in retail nurseries, thus their earnings depend on their experience, sales skill, and ambition. Some wholesale nurseries offer other benefits such as housing and/or a company vehicle in lieu of higher salaries.

Other Rewards

Anyone who enjoys working outdoors with living plant material and a wide variety of people will find a career in nursery management rewarding. To nurture your product over many years and then see it used and appreciated by others is but one of the emotional rewards of nursery work. Some people also enjoy the physical labor required and the constant challenge of working with nature and the public.

Additional information about careers in nursery management and production can be obtained from the American Nursery and Landscape Association, whose contact information is listed in Appendix B, and from individual land-grant universities in each state (found at www.nasulgc.org/About_Nasulgc/nasulgc_membership.htm).

C H A P T E R
3

CAREERS IN HORTICULTURE SERVICES

In Chapter 3, we considered horticultural careers that involve research and the actual growing and management of crops. In this chapter, we'll examine service careers whose practitioners deal directly with clients. Such jobs include floral designers, groundskeepers, landscape designers and architects, and horticultural therapists.

FLORIST, FLORAL DESIGNER

Florist and floral design careers are actually a subdivision of floriculture, and combine knowledge of plants and flowers with design techniques to produce floral arrangements, plant gifts, decorations, and related items.

Specific Work Performed

Floral designers prepare corsages, bouquets, funeral pieces, dried arrangements, wreaths, and decorations for parties, weddings, and other events. Many florists manage retail shops and are involved in the daily operation of the retail business including purchasing supplies from wholesalers, taking customer orders by phone or in person, handling complaints, making deliveries, and doing bookkeeping.

Wholesale florists purchase large quantities of flowers, plants, and floral supplies for distribution to retailers, other wholesalers in other parts of the country, or for export. They locate sources of cut flowers and supplies, prepare the flowers for storage and shipping, take orders from retailers and other wholesalers, and schedule and deliver the flowers, plants, and supplies to retail shops and others. The specific tasks performed by a floral designer and florist depend on the number of people employed and the specific services offered by the florist, ranging from single person/owner-operated shops to large floral design/events design businesses.

Job titles for those in this career field are numerous and might include the retail area sales clerk, designer, head designer, store manager, and buyer. In the wholesale area, job titles might include receiving manager, shipping manager, salesperson, sales manager, general manager, and buyer.

Specific Skills Required

Floral designers and florists must have a good sense of design and an understanding of the concepts of form, color, harmony, scale, and other design basics. Because most of the work performed by floral designers involves the public, personal interaction skills are essential. A broad knowledge of flower varieties, their seasonal availability, and their lasting qualities is required. Because floral arrangements are prepared by hand, the floral designer must have good manual dexterity.

Work Settings

Approximately one out of three floral designers is self-employed. Another eight percent work in the floral departments of grocery stores. Others are employed by miscellaneous nondurable goods merchant wholesalers, a variety of general merchandise stores, and in lawn and garden equipment and supply stores.

Almost half of all floral designers are employed in traditional service-oriented retail florist shops in both urban and rural settings. Some shops, particularly in urban areas, may be very large with many designers, a number of whom may have specialized tasks. Other shops may be very small, with a single person working as owner, manager, designer, and delivery person. These florists typically provide a range of services from customer design, potted plants, floral supplies and crafts, and delivery and wire services.

There is a continuing trend for large grocery chains and other retail concerns to offer floral products and services, thus expanding the settings where floral designers and florists can find jobs. Some florists and floral designers are employed by decorator shops that custom-make designs for weddings, parties, and other events. These shops generally do not work with walk-in trade and typically see customers only by appointment. Internet floral services also provide opportunities for floral designers.

Most floral designers and florists must stand all day and work in areas that are cool and humid to preserve flowers. Often, they will move into and out of refrigerated areas. Lifting containers of flowers and supplies weighing up to 40 pounds is not uncommon, but most materials usually weigh less than ten pounds. Adjustments can be made for workers with physical disabilities.

A 40-hour work week is common in this field, but weekend and holiday work is often the norm. Longer hours and working under pressure occur around holidays such as Mother's Day and Valentine's Day, the two most popular days to send flowers.

A background in floral design may also qualify one to teach floral design and florist shop management at the high-school, community college, and college level depending on the educational level and skills of the designer.

Florists and floral designers may also own and operate their own shops. The initial investment is high (it could be as much as $100,000 or more) and the business can be expected to turn a profit in about seven to ten years. Because florists depend on perishable materials for their work, the risk is great. Anyone contemplating opening a florist shop should have a strong knowledge of the lasting qualities of flowers, as well as a good background in business.

Employment Outlook

Employment prospects of floral designers are expected to be strong over the next several years. Job opportunities should be good because of the relatively high replacement needs in retail florists that result from comparatively low starting pay and limited opportunities for advancement.

The demand for floral designers should continue to grow as flower sales increase due to the growing population and lavishness of weddings and other special events that require floral decorations. As disposable incomes rise, more people also desire fresh flowers in their homes and offices. Increased spending on interior design is also creating a demand for stylish artificial arrangements for homes and businesses.

Opportunities should be available in grocery store and Internet floral shops as sales of floral arrangements from these outlets grow. The pre-arranged displays and gifts available in these stores appeal to consumers because of the convenience and because of prices that are lower than can be found in independent floral shops.

As mass marketers capture more of the small flower orders, independent floral shops are increasingly finding themselves under pressure to remain profitable. Many independent shops have added online ordering systems in order to compete with Internet florists. Others are trying to distinguish their services by specializing in certain areas of floral design or by combining floral design with event planning and interior design services. Some florists are also adding holiday decorating services in which they set up decorations for businesses and residences.

Few job opportunities are expected in floral wholesalers, primarily because an increasing number of shops are purchasing flowers and supplies directly from growers in order to cut costs. In addition, the growth of e-commerce in the floral industry will make it easier for retail florists to locate their own suppliers.

Advancement Opportunities

Advancement in this career field depends on the size of the business where the floral designer is employed, and most importantly on the skills of the designer. Possible areas for advancement include positions such as design supervisor, shop manager, or shop owner, or if working in an academic setting, the traditional advancement through faculty rank would be common.

Education and Training Requirements

Floral design is the only design occupation that does not require formal postsecondary training; most floral designers learn their skills on the job. Employers generally look for high-school graduates who have creativity, a flair for arranging flowers, and a desire to learn. Many florists gain their initial experience working as cashiers or delivery people in retail floral stores. The completion of formal design training, however, is an asset for floral designers, particularly those interested in advancing to chief floral designer or in opening their own businesses.

Private floral schools, vocational schools, and community colleges award certificates in floral design. These programs generally require a high-school diploma for admission and last from several weeks to one year. Floral design courses teach the basics of arranging flowers—the different types of flowers, their color and texture, cutting and taping techniques, tying bows and ribbons, proper handling and care of flowers, floral trends, and pricing.

Some floral designers also may choose to attend an associate's or bachelor's degree program at a community college or university. Some programs offer formal degrees in floral design, while others offer degrees in floriculture, horticulture, or ornamental horticulture, which can prepare students for a career in floral design. In addition to floral design courses, these programs offer classes in botany, chemistry, hydrology, microbiology, pesticides, and soil management.

Since many floral designers manage their own business, additional courses in business, accounting, marketing, and computer technology can be helpful.

The American Institute of Floral Designers (AIFD) offers an accreditation examination as an indication of professional achievement in floral design. The exam consists of a written part covering floral terminology and an onsite floral-arranging test in which candidates have four hours to complete five floral designs. The five categories of floral designs are funeral tributes, table arrangements, wedding arrangements, wearable flowers, and one category of the candidate's choosing. (See Appendix A for AIFD contact information.)

Summer jobs, internships, and cooperative education can provide important work experience. Positions exist with retail florists, greenhouses, wholesale florists, and floral supply businesses.

Other Personal Qualifications

Floral designers and florists need a good sense of color and design as well as physical stamina and manual dexterity. Good verbal communication skills are essential for interacting with both suppliers and customers. They need to be creative, precise, and careful in their work. Florists must be aware of the necessity of meeting deadlines and public satisfaction with their work. Although some floral designers may work alone with little interaction, most benefit from good interpersonal skills considering the amount of interaction with colleagues and the public. In the present job market, computer skills are helpful for placing supply orders, maintaining inventory records, and doing bookkeeping.

Salary and Other Benefits

Salaries for floral designers and florists depend on the level of skill and education of the individual, the size of the business, the geographic area of employment, and the economy of the area. For instance, entry-level floral designers and florists may be paid only minimum wage in small rural shops. However, as skills and creativity are proven, the pay rate will rise on an hourly basis.

In 2004, median annual earnings for salaried floral designers were $20,450. The middle 50 percent earned between $16,670 and $25,610. The lowest 10 percent earned less than $14,360, while the highest 10 percent earned more than $32,370. Median annual earnings were $22,520 in grocery stores and $20,110 in florists. Salaries can be significantly higher for highly skilled designers employed in busy urban areas or for those who work at highly specialized shops.

Professionals trained in floral design and who held master's and Ph.D. degrees generally were employed in colleges and universities. These people typically had horticultural training and skills beyond floral design, in areas such as the growing of cut flowers and other floral crops and horticultural business management. Salaries were variable depending on the college or university and the academic rank held by the person. Average salaries ranged from a high of $90,900 for a full professor to a low of $22,900 for an instructor.

Other Rewards

Probably the most obvious nonmonetary benefit of a career as a floral designer and florist is the opportunity to work with flowers, creating a beautiful product that brings pleasure to the person who receives it. Most florists must also interact with the public daily, an obvious plus for anyone who does not like to work alone. Floral designers and florists are generally active; thus, if you wouldn't be happy with a job that requires you to sit all day at a desk, then a career as a florist and floral designer may prove rewarding.

LANDSCAPE CONTRACTOR, GROUNDSKEEPER, GARDENER

Landscape contractors, groundskeepers, and gardeners maintain and install landscapes at both residential and commercial sites. Landscape contractors are independent businesses contracted by the homeowner or business client. They offer a wide range of services, presenting a diverse array of career opportunities. Groundskeepers and gardeners are typically staff personnel employed directly by businesses and homeowners to maintain and oversee their personal landscapes. Golf course managers and related personnel are included in this career category although their qualifications, training, and work are more specific and unique.

Job titles in this career field are as diverse as the field itself and include salesperson, operations manager, project manager or coordinator, construction foreman, maintenance foreman, construction and maintenance coordinator, arborist, grounds foreman, golf course superintendent, turf manager, and head gardener.

Specific Work Performed

Those who work in this career field perform a very broad range of work, which varies according to the business type or setting. In general, landscape contractors, groundskeepers, gardeners, and golf course personnel all tend lawns, shrubs, trees, and flowers. Each may also be expected to be involved in planning and planting landscapes and building and maintaining related structures.

Landscape contractors offering maintenance services cut, trim, water, fertilize, and rake lawns; water, fertilize, prune, and spray trees, shrubs, and flowers; and remove litter and snow. Those involved in the installation aspect of the field buy plants from wholesale nurseries, design landscapes and gardens, grade and prepare sites for planting, and install trees, shrubs, ground covers, and turf, and related structures such as retaining walls, fences, patios, and even driveways, pools, and irrigation systems. Some landscape contractors act as general contractors, subcontracting some of the building portions of the work. Landscape contractors involved in installation often keep landscape designers or landscape architects and estimators on staff. These businesses are sometimes referred to as design/build firms.

Some landscape contractors offer design, installation, and maintenance services to their clients, but others limit their work to one or two of these services. The size of the contracting firm and the services offered determine the types of personnel employed and the specific tasks performed. Almost all landscape contractors employ salespersons, field foremen, and supervisors, in addition to laborers who perform the specific tasks associated with maintenance and construction.

Groundskeepers and gardeners employed on staff by businesses and individuals take care of lawns, trees, shrubs, and flowers. Specific job duties include mowing, trimming, and watering lawns and plants, pruning trees and shrubs, raking leaves, and removing litter and snow. They also apply fertilizers, herbicides, insecticides, and fungicides. In addition, they install or oversee the installation of trees, shrubs, and other plant and related landscape materials. Some groundskeepers and gardeners supervise other personnel who perform the specific tasks mentioned.

Golf course personnel take care of the trees, shrubs, ground covers, and flowers on private and public golf courses. Their primary concern is the maintenance of the turf areas of the course, including the rough, the fairways, and the greens areas. They must mow, water, fertilize, and otherwise intensively care for the landscape and turf areas of the course for high-quality and optimal playing conditions. Most golf course workers must apply pesticides to provide a turf that meets both aesthetic and playing standards, and must do so remaining aware of the environmental concerns associated with the application of chemicals, especially those as intensively needed on golf courses. Golf course personnel are also involved in planning, selecting, and planting trees, shrubs, and other plant materials, and operate and maintain irrigation systems.

Employment in these fields is sometimes seasonal, especially for field personnel. Many employers attempt to keep supervisory personnel year-round by including winter services such as snow removal or holiday decorating and by emphasizing sales, planning, or other areas during the slow season.

Specific Skills Required

Given the variety of tasks performed in these fields, the specific skills required are highly varied and range from the ability to perform routine tasks such as pruning shrubs with a hand pruner to supervising the installation of large trees to managing the overall landscape operations of a large public facility. The skills required for persons employed in these areas might include the ability to properly operate lawn mowers, trimmers, hand pruners, forklifts, backhoes, and even larger equipment such as tree spades. Most workers in this field should be able to identify trees, shrubs, and other plants commonly found in landscapes and have an understanding of the maintenance requirements of these plants, including watering, fertilizing, pruning, and pest control requirements.

Employees involved in the installation of plants must know the proper techniques for digging, loading, transporting, and planting the trees, shrubs, and other plants. They should know how to prepare a site for planting and be familiar with grading, drainage, and soils. Supervisory personnel in the field must have excellent leadership and written and oral communication skills, possibly including the mastery of a foreign language.

Supervisors must have training and experience in doing the various construction or maintenance tasks they oversee, and must be able to train workers in the procedures to be used and the equipment to be operated. Sales and business management skills are very beneficial. Anyone directly responsible for the application of pesticides must be licensed through a state examination process.

Work Settings

People employed in these fields generally spend a lot of their work time outdoors at the sites being planted and maintained, often in inclement weather. Supervisory, sales, and design personnel work indoors as well. Job settings include private residences, office parks and shopping centers, commercial and public sites, parks, plant nurseries, zoos, cemeteries, historic landscape sites, public and private gardens, hotels and resorts, arboreta and botanical gardens, and public and private clubs and golf courses. Work hours are long and often irregular, dictated by weather, client demands, available personnel, and work schedules. Six-day work weeks are customary, especially in spring, and physical and mental stress are common during periods of heavy work load.

Employment Outlook

Those interested in grounds maintenance occupations should find plentiful job opportunities in the future. Demand for their services is growing, and because wages for beginners are low and the work is physically demanding, many employers have difficulty attracting enough workers to fill all openings, creating very good job opportunities. In addition, high turnover should generate a large number of job openings, including those at the supervisory and managerial level.

More workers should also be needed to keep up with the increasing demand by lawn care and landscaping companies. Expected growth in the construction of all types of buildings, from office complexes to shopping malls and residential housing, plus more highways and parks, should increase the demand for grounds maintenance workers. In addition, the upkeep and renovation of existing landscaping and grounds are continuing sources of work for grounds maintenance employees. Owners of many buildings and facilities recognize the importance of "curb appeal" in attracting business and maintaining the value of the property and are expected to use grounds maintenance services more extensively to maintain and upgrade their properties. Grounds maintenance employees working for state and local governments, however, may face budget cuts, which could affect hiring.

Homeowners are a growing source of demand for grounds maintenance workers. Many know that a nice yard will increase the property's value. In addition, there is a growing interest by homeowners in their backyards, as well as a desire to make the yards more attractive for outdoor entertaining. With many newer homes having more and bigger windows overlooking the yard, it becomes increasingly important to maintain and beautify the grounds. Also, as the population ages, more elderly homeowners will require lawn care services to help maintain their yards.

Job opportunities for tree trimmers and pruners should also increase as utility companies step up the pruning of trees around electric lines to prevent power outages. Additionally, tree trimmers and pruners are needed to help combat infestations caused by new species of insects from other countries. Ash trees in Michigan, for example, have been especially hurt by a pest from China.

Job opportunities for nonseasonal work are more numerous in regions with temperate climates, where landscaping and lawn services are required all year. However, opportunities may vary with local economic conditions.

Advancement Opportunities

Advancement opportunities in these fields depend on ambition, education, and experience. Many people begin working in these occupations as laborers with very small companies. As they gain experience and/or more education and training, however, workers can be promoted to positions such as foreman, supervisor, or manager, while some may make the jump to larger businesses or operations. Other employees may move into sales and design positions as their education and skills improve. Some workers, of course, eventually begin their own businesses. Like many of the career fields in horticulture, advancement is limited only by the individual's ability to perform the job.

Education and Training Requirements

Many groundskeepers, gardeners, and employees of landscape contractors do not have any formal education in horticulture, relying completely on prior work experience for entry into, or advancement in, the profession. However, as competition for positions increases, a two-year associate's degree may become the minimum educational level for employment. Those who hope to advance to supervisory and management positions should consider a bachelor's degree.

High-school courses in science, mathematics, art, drafting, business, and horticulture are a good way to prepare for this field. Students considering a college degree program should take the recommended college preparatory courses, as well as additional courses in business, foreign language, computer skills, chemistry, and biology as support courses. Two-year degrees preparing you for these fields should contain core courses required of all students, plus specific courses in horticulture depending on the areas in which you wish to work. Typically, students take courses in general horticulture, plant materials, soils, chemistry, landscape maintenance and management, landscape construction, business management, and pest identification and management.

On-the-job training is common in these fields, and experience is invaluable for entry into, and advancement in, these career fields. However, education beyond high school should give you greater access and opportunities, helping you move into supervisory and management positions more quickly.

Summer and part-time jobs, internships, and cooperative education experience are important to help get the experience needed to be successful in these fields. Positions exist with landscape contractors, grounds departments in the public and private sectors, public and private gardens, and arboreta and botanical gardens. Some arboreta and botanical gardens even offer certificate programs that help prepare the student for entry into these fields.

Those who plan to enter the field of golf course management need a minimum of a bachelor's degree. Certification may provide some edge for entry into, and advancement in, the profession, but it is not required. Course work will likely concentrate on turf management, soils, irrigation, pest control practices, and grounds management and maintenance. Again, any related experience prior to entering the workforce is highly desirable. Complete information about golf course management is available from the Golf Course Superintendents Association of America, whose contact information is listed in Appendix B.

Most states require certification for workers who apply pesticides. Certification requirements vary, but usually include passing a test on the proper and safe use and disposal of insecticides, herbicides, and fungicides. Some states require that landscape contractors be licensed.

The Professional Grounds Management Society (PGMS) offers certification to grounds managers who have a combination of eight years of experience and formal education beyond high school and who pass an examination covering subjects such as equipment management, personnel management, environmental issues, turf care, ornamentals, and circulatory systems. The PGMS also offers certification to groundskeepers who have a high-school diploma or the equivalent, plus two years of experience in the grounds maintenance field. (See Appendix B for contact information.)

The Professional Landcare Network (PLANET) offers the designations Certified Landscape Professional (Exterior and Interior) and Certified Landscape Technician (Exterior or Interior) to those who meet established education and experience standards and who pass a specific examination. The hands-on test for technicians covers areas such as the operation of maintenance equipment and the installation of plants by reading a plan. A written safety test also is administered. PLANET also offers the designations Certified Turfgrass Professional (CTP) and Certified Ornamental Landscape Professional (COLP), which require written exams. (See Appendix B for contact information.)

Other Personal Qualifications

People who are successful in the fields of landscape contracting, groundskeeping, gardening, and golf course management are organized, artistic, and detail oriented. They have an interest in working with a diverse group of people from a variety of educational and social backgrounds. They also like to work with the public. Obviously an interest in nature is important for success in these fields as well.

Physical demands in this career field include the ability to lift heavy objects weighing 100 or more pounds, and unrestricted use of the legs, arms, and hands for kneeling, reaching, stooping, and climbing. Because of the diversity of tasks in these fields, adjustments can be made for persons with physical disabilities.

Salary

Data on salaries in these fields is limited and earnings will vary widely depending on the position held, the type and size of the employer, education, and experience. Laborers and entry-level employees without experience or education may be paid minimum wage. Hourly wages are common even for experienced employees and range from $10 to $17. Salaried employees may be paid from $20,000 to $35,000 per year, but experienced and well-educated supervisory and management level workers can earn $50,000 per year or more.

Other Rewards

Anyone who enjoys working outdoors and with a diverse group of people will find these career fields rewarding. The joy of working with nature and creating and maintaining aesthetic and useful outdoor environments for others to enjoy is appealing to many people. The physical labor and mental challenges provided by this work can be highly rewarding also.

LANDSCAPE DESIGNER

The practice of landscape design dates back thousands of years to ancient Egypt. The early pharaohs commissioned the design of formal enclosed gardens, leading to a highly developed knowledge of garden design that quickly spread throughout the civilized world. Many of the theories and practices of these early garden designers persist even to this day. In fact, it was those early Egyptian designers who developed the idea of using irrigation systems (canals) in order to build lush and productive gardens in a desert climate.

Landscape design is sometimes confused with the fields of garden design and landscape architecture. Although these career fields are similar in some respects, they are each distinct in the actual tasks performed and in the education and training they require.

To state the obvious, garden designers design gardens. Some may have knowledge of structural elements, and most have come to the profession out of a personal love for plants and gardens. Many aren't formally trained in drafting and design, nor do they necessarily have any formal education in horticulture. In general, garden designers are self-taught through experience or have had some informal training through local plant societies, arboreta, and botanical gardens. Many do wonderful design work and are very successful in their profession due to an innate talent and years of experience working with their own gardens.

Landscape designers are professionals who plan, design, and/or construct exterior spaces using plant materials as well as paving and building materials. In comparison to garden designers, landscape designers generally have a more formal education in drafting and design techniques, horticulture, and the correct and effective use of structural elements such as paving materials, walls, decking, and fencing.

Landscape architects, in contrast to both garden designers and landscape designers, generally work on a larger scale with municipal or commercial projects, although many have residential clients. Landscape architects are formally educated and trained at the college or university level or have had a minimum of ten years' experience and have passed a rigorous state exam. Although some landscape architects do have a strong background in horticulture, many are trained primarily in "hardscaping"—that is, designing drainage systems, driveways, parking lots, retaining walls, or determining the locations of buildings or other structures.

Each of these professions contributes uniquely to the design and installation of many landscape elements, from flowers to major roadways, and each requires a specific level of education and training for career success. Because of the more professional nature of the occupations of landscape designer and architect, we will explore these careers in more detail in the following pages.

Specific Work Performed

Landscape designers typically work on residential or small commercial projects. They plan and design exterior and interior spaces to satisfy the needs and desires of the client in an aesthetically pleasing way. The designer consults with the clients to evaluate their needs and the restrictions and possibilities of the site. Next, the designer prepares drawings for the project showing the locations of plants, walkways, walls, fences, water features, and other design elements. Specific plant materials and building materials and methods are determined. Some designers estimate the cost of the work to be performed and may act as consultant to the client with a contractor hired to install the plan. Qualified designers may also act as general contractors and oversee the entire project.

Specific Skills Required

Landscape designers must have an excellent sense of design, including an understanding of the concepts of form, color, harmony, scale, and other design basics. They must be imaginative and artistic. Strong drafting skills are required for the production of accurate plan view, elevation, and detail drawings. In order to solve a variety of design problems and challenges, designers need a thorough knowledge of a wide variety of plant materials, their growth habits and requirements, as well as their proper usage.

Landscape designers typically are knowledgeable about landscape styles such as formal, informal, naturalistic, Japanese, and so on, and have a grasp of landscape design history. They are also familiar with soils and their properties, drainage, and building materials and methods. In addition, they have an understanding of climate and weather and the effects of rainfall, sun, temperature, and wind on plant materials and other landscape materials. Other areas of knowledge include information on local zoning and building codes, as well as environmental regulations.

An interest in problem solving, the ability to make decisions, and a concern for accuracy and detail is helpful for the successful landscape designer. A good understanding of mathematics and geometry, business management, and public relations is also very useful. Because landscape designers must interact with a wide variety of people—including homeowners, contractors, nurserymen, field laborers, and public officials—strong interpersonal skills are essential. Good computer skills are also a necessity, including knowledge of the more common software programs, as well as some familiarity with CAD programs (computer-aided design).

Work Settings

The work settings for landscape designers can be as diverse as the tasks performed and skills required for this career choice. Landscape designers typically divide their time between outdoor and indoor activities. Outdoor activities can include evaluating sites and overseeing projects, and choosing plants and hardscape materials, while indoor office duties may involve researching and preparing designs and drawings and a variety of support documents, as well as meeting with clients.

Office settings can range from sparse to elaborate, depending on the size of the company. Some designers work alone, sometimes out of their homes; others work for landscape contractors and nurseries ranging from small to large businesses; and some designers work on staff, or as consultants, for architecture, landscape architecture, engineering, or land planning firms, academic institutions, or government agencies. The size and type of firm or office where the designer is employed will determine the exact nature of the work setting. However, in general, work settings for designers are comfortable and well equipped. Many landscape designers spend most of their time at drafting tables, while others may work primarily with sales and client contact. Some draw their own plans, but others may have their drawings done by drafting personnel. Frequent interaction with the public and those in related and support professions is obviously necessary; thus, persons preferring to work alone should reconsider this profession as a career choice.

Landscape designers also work in academic institutions and in government agencies. Designers with a master's degree and experience frequently teach design and related courses at junior and community colleges, and occasionally at the college and university level depending on the degrees earned. In this career field, experience and success in the profession can substitute for a formal education. Others may be employed on college and university campuses as a part of the buildings and grounds department. Landscape designers also work for local, state, and federal agencies in areas such as urban planning, parks and recreation, environmental planning and protection, and transportation. Public and private arboreta and botanical gardens, cemeteries, and golf courses frequently employ landscape designers on staff or as consultants.

Landscape designers rarely work regular hours, the length of the work day being determined by project deadlines. Weekend work is also common, because this is often the only time clients can meet with the designer. Work is most intense during spring and summer in geographic areas where project installation can be inhibited by winter weather, but landscape designers will typically concentrate on sales for the upcoming season during the winter months.

Although you might think of landscape designers as working only on outdoor spaces, the interior landscaping business is a growing area of landscape design. Designers involved in this area work primarily on large-scale commercial installations for shopping malls, office parks, hotels, convention and meeting centers, and other similar interior sites. The specific skills, education, and training required for an interior landscaping career are the same as those for other designers. However, the interior designer must be knowledgeable about tropical and seasonal flowering plant materials, interior irrigation systems, containers, and the appropriate climatic conditions necessary for survival of interior plants. Some familiarity with the impact of interior environments and public traffic on the growth of interior plants is also needed. This is a fairly young subdivision of the landscape design field and an ambitious and talented interior landscape designer could be very successful.

The work settings where you might find a landscape designer are highly diverse and varied in size, clientele, and the nature of the work performed. Opportunities are almost unlimited for a person interested in this career field to find a place where he or she can be happy and successful.

Employment Outlook

The employment outlook for landscape designers is strong. Because designers are able to work in a wide variety of settings, ranging from self-employed independent designers to hourly or salaried consultants in large architecture and engineering firms in both the private and public sectors, the number of positions available should remain good. With increasing consumer interest in the environment and in improving the exterior and interior spaces in which we work and play, qualified landscape designers will remain in demand.

Any downturn in the economy may reduce employment opportunities in firms targeting lower-end projects such as middle-income residences or small-scale commercial projects, but employment in firms targeting high-end residential and large commercial projects should remain strong. As populations continue to increase and shift from urban to suburban, the importance of usable and aesthetically pleasant exterior and interior spaces should grow. The general public has accepted that quality landscaping adds to the value of both residential and commercial properties, thus the need for creative and qualified landscape designers should increase.

Advancement Opportunities

Advancement in the landscape design profession depends on the size and type of office, as well as the designer's own experience and ambition. Typically, a designer may begin by drafting the work of more experienced designers or architects. Advancement may result in the designer being asked to contribute design solutions or suggestions to a project or he or she may be given specific sections of the project to design or supervise.

As skills improve, the landscape designer may be given entire projects to oversee, the size or importance of the projects eventually increasing. This scenario may be described as advancing from draftsperson to junior designer to senior designer to project supervisor or director.

Other designers may advance into sales or staff management positions, if this is the career path they have chosen. Some may begin their careers self-employed and advance into staff positions with private or public employers, while others often gain experience working for companies or contractors before advancing into private practice.

Education and Training Requirements

Formal education and training in landscape design are essential for success. High-school courses in biology, mathematics and geometry, chemistry, drafting, design and art, and computer skills can be highly beneficial for students who hope to enter this profession. Summer and after-school work with landscape contractors, nurseries, public and private gardens, grounds departments, or design firms can provide additional training and exposure to the diverse activities associated with landscape design.

Some botanical gardens, such as Longwood Gardens in Pennsylvania and the New York Botanical Garden, offer certificate programs in landscape design. Generally, an associate or bachelor's degree is the minimum education needed for success in landscape design. In most schools, these degrees can be obtained in the horticulture department.

Some colleges and universities offer master's programs in landscape design. Specific course work depends on the school attended, but generally (beyond the core requirements) additional courses in design theory and principles, design history, landscape plant materials, drafting, surveying, and computer-aided design may be needed. Landscape designers also may take courses in soil science, plant pest identification and control, landscape construction, landscape maintenance, and art. Classes in technical writing, public speaking, and business management are also helpful.

Summer jobs, internships, and cooperative education can provide important work experience for students. Positions exist in private businesses, government agencies, and on college campuses.

In most landscape design programs, a senior design project is required for graduation. This type of project can help the student decide on a particular area of design he or she will like best or least, and it can give the student invaluable insight into how the profession works. If the program does not require a design project for graduation, students are encouraged to volunteer to work with a faculty member on one of his or her professional projects. Students should take advantage of all opportunities to gain experience, especially those involving supervision by a qualified professional landscape designer.

Though landscape designers are not required to be licensed, today there is increasing interest in their certification. Anyone can legally call himself or herself a landscape designer regardless of actual education, training, or skill. Therefore, in the interest of improving the professional image of its members, as well as protecting the public from poorly trained or untrained designers, the profession has established a certification program through the Association of Professional Landscape Designers (APLD). A landscape designer with a minimum of two years' professional experience can be certified by the APLD. The designer must also submit landscape plans that are evaluated by a committee to ensure that they meet current professional standards. Some state nursery and landscape associations offer landscape design certification through their nursery certification programs. Anyone considering landscape design as a profession should also consider the necessity of professional certification. For additional information regarding certification, contact APLD or your local cooperative extension service. (Contact information for both organizations is listed in Appendix B.)

Other Personal Qualifications

Landscape designers are typically curious and creative people who can visualize solutions to problems. They usually do not like routine and are able to apply their education and training based on past successes and failures in order to solve new problems. Successful designers are rarely afraid of expressing their opinions, but they are conscious of the need for strong public relations skills. Physical stamina is needed for the long hours of design and drafting required in most practices. However, no minimum strength or other physical requirements are necessary. An appreciation of nature and the environment is basic to this profession, as is the desire to make our surroundings a more pleasant place in which to work and play.

Salary and Other Benefits

Statistics on the salaries of landscape designers are limited because of the diverse work settings and skill requirements within the field, as well as the relative youth of landscape design as a profession. In general, landscape designers employed by garden centers are paid a lower salary plus commission, while those who work for design firms earn a straight salary. Average starting salaries in landscape design range from $25,000 to $32,000 per year. Salaries vary depending on a designer's experience and education.

Salaries for college teachers depend upon rank, the type of institution, and the geographic area. According to a 2004–05 survey by the American Association of University Professors, salaries for full-time faculty averaged $68,505. By rank, the average was $91,548 for professors, $65,113 for associate professors, $54,571 for assistant professors, $39,899 for instructors, and $45,647 for lecturers. It should be noted that designers holding positions of associate and full professor typically held master's or doctorate degrees and had many years of successful experience in the practice of landscape design.

Landscape design professionals are paid a broad range of salaries dependent on education level, the type of work performed, the size or type of employer, and the geographic location. If salary is a major concern when choosing this career field, the student should seek to obtain the highest level of education possible and to develop his or her design skills to assure the compensation desired.

Other Rewards

The greatest reward for the successful landscape designer is the satisfaction of creating something that makes people's lives more pleasant. Other non-monetary rewards include working with nature or with a variety of people who might be your clients or colleagues. Although most landscape designers do not become wealthy, they are just as self-sufficient as members of other design professions like architects, interior planners and designers, engineers, and landscape architects.

LANDSCAPE ARCHITECT

Although a relatively new profession, landscape architecture can trace its origins back thousands of years to the early practice of garden design or landscape design in ancient Egypt. From the early Egyptians, this knowledge of garden design spread and increased throughout the Western world. In the Far East, early civilizations also developed the concept of landscape or garden design, although influenced by different principles and theories growing out of their unique culture. Out of the more ancient practices of landscape and garden design, as well as horticulture and agriculture, the practice of landscape architecture has gained the professional status it enjoys today.

Currently, it is an evolving career field. While originally landscape architects applied their skills to the design of gardens and formal estates and parks, these days professional landscape architects are also concerned with the preservation of our dwindling natural resources and the creation of beautiful and useful outdoor environments for everyone.

Specific Work Performed

Landscape architects design outdoor spaces that are functional and beautiful while taking into account the impact of the design on the natural environment. They work on projects ranging from home gardens to larger, more complex endeavors such as public parks or land preservation sites. Most specialize in a particular area of design, which includes, but is not limited to, small residential design; estate design; small commercial design; the design of large-scale commercial sites such as office parks, shopping centers, housing developments, cemeteries, athletic fields and golf courses, botanical gardens and arboreta, and university campuses; public park, zoo, and recreational area design; or highway planting design. Today, some landscape architects specialize in environmental studies and preservation, land reclamation, or the design of forest preserves and nature conservation sites. A few, meanwhile, focus their careers on landscape architecture history, writing, or teaching.

Specific work performed by landscape architects depends on the area or areas of specialization chosen, but almost all practicing architects are involved in tasks related to clients and the projects they hope to build on a specific piece of land. Exceptions might occur if the landscape architect teaches or works for a government agency.

After studying and analyzing the site, landscape architects prepare a preliminary design. To account for the needs of the client as well as the conditions at the site, they frequently make changes before a final design is approved. They also take into account any local, State, or Federal regulations, such as those protecting wetlands or historic resources. In preparing designs, computer-aided design (CAD) has become an essential tool for most landscape architects. Most also use video simulation to help clients envision proposed ideas and plans. For larger-scale site planning, landscape architects also use geographic information systems technology—that is, a computer mapping system.

Throughout all phases of the planning and design, landscape architects consult with other professionals—such as civil engineers, hydrologists, or architects—involved in the project. Once the design is complete, they prepare a proposal for the client. They produce detailed plans of the site, including written reports, sketches, models, photographs, land-use studies, and cost estimates, and then submit them for approval by the client and by regulatory agencies. When the plans are approved, landscape architects prepare working drawings showing all existing and proposed features. They also outline in detail the methods of construction, and draw up a list of necessary materials. Afterward, landscape architects mainly monitor the implementation of their design, with general contractors or landscape contractors usually directing the actual construction of the site and the installation of plantings.

Landscape architects who work for government agencies do site and landscape design for government buildings, parks, and other public lands, as well as park and recreation planning in national parks and forests. In addition, they prepare environmental impact statements and studies on issues such as public land-use planning. Some restore degraded land, such as mines or landfills, while other landscape architects use their skills for traffic-calming, the "art" of slowing traffic down through the use of traffic design, enhancement of the physical environment, and a greater attention to aesthetics.

In small landscape architecture practices, each landscape architect would be expected to perform most or all of these tasks, but in larger offices the landscape architect may function in a more supervisory and administrative capacity, assigning each of the aforementioned tasks to draftsmen, estimators, project coordinators, surveyors, office managers, and other assistants.

Most successful landscape architects work on several projects simultaneously, with each project likely at a different stage of development. Thus, landscape architects are constantly involved in the various aspects of existing contracts, and during this time both analyze new projects and contact potential clients.

It is important to emphasize that the specific work performed by landscape architects is more than the design and drawing of gardens. Although this is a part of the work, it is only a very small part. Landscape architects are also involved daily in the construction, engineering, political, and business details of the profession. But this diversity of work tasks is a major aspect of this career field, and is in great part what makes it so appealing to those who choose landscape architecture as their profession.

Specific Skills Required

Landscape architects must be imaginative and artistic people with an appreciation for nature who can use their talents and skills to solve problems. They must possess an excellent grasp of the principles of design, including form, scale, color, and other design basics. At the same time, landscape architects must be highly organized in order to juggle the variety of tasks performed on a daily basis. They must also be well informed regarding local zoning laws, building codes, and environmental regulations.

Excellent mathematical and engineering skills, plus an expertise in surveying and drafting are necessary for the production and implementation of accurate plan view, elevation, and detail drawings. Knowledge of landscape design history and the variety of landscape styles available, coupled with a good understanding of both plant materials and hardscape materials—and their proper usage—is essential. Landscape architects must also have a thorough understanding of soils, drainage, and the effects of time, climate, and weather on the plants and building materials used to create outdoor spaces. Today, not only must landscape architects have good computer skills in order to prepare plans (CAD) or other documents, they need excellent written and verbal communication skills, and should be able to make decisions and accept responsibility, and interact successfully with the wide variety of people they encounter, from clients and contractors to politicians and civic committees.

As explained earlier, every landscape architect may not need all of these specific skills to be successful in their profession. The great variety of work in landscape architecture offers opportunities for a wide range of people with many diverse skills.

Work Settings

The nature of their work requires that landscape architects divide their time between indoor and outdoor duties. Outdoors, they are involved in the site analysis and surveying aspects of the work. Indoors, they may work at drafting tables preparing drawings and other documents, or they may work in more typical office surroundings meeting with clients, discussing and presenting design ideas, or preparing proposals and contracts. Some landscape architects work alone or with a few assistants in small offices, but others may work in large architecture, engineering, or landscape architecture firms with numerous assistants and highly technical and modern equipment. In general, work settings for landscape architects are comfortable and well equipped, but the size and type of firm where the landscape architect is employed will determine the exact nature of the work setting.

Some landscape architects work for botanical gardens and arboreta, while others may be employed by university and college grounds departments. Historical organizations such as Colonial Williamsburg, cemeteries, resorts and theme parks, and golf courses may also have landscape architects on staff.

Landscape architects also work teaching in academic settings or for federal agencies such as the National Forest Service, the National Park Service, and the U.S. Corps of Engineers. These agencies employ landscape architects for the planning, management, and preservation of large tracts of public land. Similar positions exist at the state and local level in highway departments, parks departments, and planning departments. Some landscape architects even work overseas, especially in developing or "third-world" countries for foreign governments and private architectural and engineering firms engaged in the design and construction of new or renovated outdoor environments for those populations.

Typically, landscape architects work long and irregular hours, and weekend work is common. Some prefer working for institutions and government agencies where work hours are somewhat more regular. The work load may be somewhat seasonal, especially in smaller firms and in geographic locations where weather and climate may interfere with outdoor work. Most landscape architecture professionals concentrate on sales or long-term projects during these periods.

Many landscape architects are self-employed because start-up costs, after an initial investment in CAD software, are relatively low. Self-discipline, business acumen, and good marketing skills are important qualities for those who choose to open their own business. Even with these qualities, however, some may struggle while building a client base.

Those with landscape architecture training also qualify for jobs closely related to landscape architecture, and may, after gaining some experience, become construction supervisors, land or environmental planners, or landscape consultants.

Employment Outlook

Employment of landscape architects is expected to grow significantly over the next several years. A steady increase in the planning and development of new residential, commercial, and other types of construction should lead to employment opportunities in the field. With land costs rising and the public desiring more beautiful spaces, the importance of good site planning and landscape design is growing. Demand for landscape architecture services should also increase due to new demands to manage stormwater run-off in both existing and new landscapes, combined with the growing need to manage water resources in the western states.

New construction also is increasingly contingent upon compliance with environmental regulations, zoning laws, and water restrictions, which should spur demand for landscape architects to help plan sites that meet these requirements and integrate new structures with the natural environment in the least disruptive way. Landscape architects will also likely be increasingly involved in preserving and restoring wetlands and other environmentally sensitive sites.

Continuation of the Transportation Equity Act for the 21st century is also expected to spur employment for landscape architects, particularly through state and local governments. This act, known as TEA-21, provides funds for surface transportation and transit programs, such as interstate highway construction and maintenance, and environment-friendly pedestrian and bicycle trails.

In addition to the work related to new development and construction, landscape architects are expected to be involved in historic preservation, land reclamation, and the refurbishment of existing sites. They are also likely to do more residential design work as households continue to spend more money on landscaping than in the past.

Advancement Opportunities

Most new employees in this field, particularly those in states where licensure is required, may be called "apprentices" or "intern landscape architects" until they become licensed. Their duties vary depending on the type and size of the employing firm. They may do project research or prepare working drawings, construction documents, or base maps of the area to be landscaped. Some are allowed to participate in the actual design of a project. However, interns must perform all work under the supervision of a licensed landscape architect. Additionally, all drawings and specifications must be signed and sealed by the licensed landscape architect, who takes legal responsibility for the work.

After gaining experience and becoming licensed, landscape architects usually can carry a design through all stages of development. After several years, they may become project managers, taking on the responsibility for meeting schedules and budgets, in addition to overseeing the project design. Later, they may become associates or partners of a firm, with a proprietary interest in the business.

Education and Training Requirements

A bachelor's or master's degree in landscape architecture usually is necessary for entry into the profession. A bachelor's degree in landscape architecture takes four or five years to complete. There are also two types of accredited master's degree programs. The more common is a three-year first professional degree program designed for students with an under-graduate degree in another discipline. The second is a two-year second professional degree program for students who have a bachelor's degree in landscape architecture and who wish to teach or specialize in some aspect of landscape architecture, such as regional planning or golf course design.

Approximately 60 colleges and universities offer undergraduate and graduate programs in landscape architecture that are accredited by the Landscape Architecture Accreditation Board of the American Society of Landscape Architects (ASLA). College courses required in these programs usually include technical subjects such as surveying, landscape design and construction, landscape ecology, site design, and urban and regional planning. Other classes include history of landscape architecture, plant and soil science, geology, professional practice, and general management. The design studio is another important aspect of many landscape architecture curriculums. Whenever possible, students are assigned real projects, providing them with valuable hands-on experience. While working on these projects, students become more proficient in the use of computer-aided design, geographic information systems, and video simulation.

Forty-seven states require landscape architects to be licensed or regis-tered. Licensing is based on the Landscape Architect Registration Examination (LARE), sponsored by the Council of Landscape Architectural Registration Boards, and is administered in two portions: graphic and multiple choice. Each portion of the testing is conducted over two days. Admission to the exam usually requires a degree from an accredited school plus one to four years of work experience under the supervision of a registered landscape architect, although standards vary from state to state. Currently, 14 states require that a state examination be passed in addition to the LARE to satisfy registration requirements. State examinations, which usually are one hour in length and are completed at the end of the LARE, focus on laws, environmental regulations, plants, soils, climate, and any other character-istics unique to the state.

Because state requirements for licensure are not uniform, landscape architects may not find it easy to transfer their registration from one state to another. However, those who meet the national standards of graduating from an accredited program—serving three years of internship under the supervision of a registered landscape architect, and passing the LARE—can satisfy requirements in most states. Through this means, a landscape architect can obtain certification from the Council of Landscape Architectural Registration Boards, and so gain reciprocity (the right to work) in other states.

Candidates for entry positions in the federal government don't need to be licensed, but they should still have a bachelor's or master's degree in landscape architecture.

Other Personal Qualifications

Those planning a career in landscape architecture should appreciate nature, enjoy working with their hands, and possess strong analytical skills. Creative vision and artistic talent are also desirable qualities. In addition, good oral communication skills are essential since landscape architects must be able to convey their ideas to other professionals and clients, and make presentations before large groups. Strong writing skills are also valuable, as is knowledge of computer applications of all kinds, including word processing, desktop publishing, and spreadsheet programs. Landscape architects use these tools to develop presentations, proposals, reports, and land impact studies for clients, colleagues, and superiors. The ability to draft and design using CAD software is essential.

Many employers recommend that prospective landscape architects complete at least one summer internship with a landscape architecture firm in order to gain an understanding of the day-to-day operations of a small business, including how to win clients, generate fees, and work within a budget.

Salary

Compensation for landscape architects depends on the educational level and skills of the architect, the types of projects on which he or she works, and the size of the firm in which he or she is employed. Landscape architects employed by government agencies typically are paid more than those in the private sector.

In 2004, median annual earnings for all landscape architects were $53,120. Those with less than five years' experience earned an average of $41,803. Landscape architects with 20 years' experience earned $80,273, while those with 40 years' experience earned an average salary of $97,564.

Architectural, engineering, and related services employed more landscape architects than any other group of industries, generating median annual earnings of $51,670.

Landscape architects who teach in colleges and universities earn the average salaries for their professional ranks, as discussed earlier. Many in academia supplement their teaching salaries with consulting and private practice. Thus, salaries for landscape architects are limited only by the skills and initiative of the individual landscape architect.

In 2005, the average annual salary for all landscape architects in the federal government in nonsupervisory, supervisory, and managerial positions was $74,508.

Other Rewards

Rewards other than salary associated with a career in landscape architecture include the admiration of others, both peers and clients, for the skill of your work. Working outdoors can be rewarding and interacting with a diverse range of people can add to the pleasure of the work. If you're interested in traveling, a landscape architecture career can provide such opportunities. Perhaps the greatest reward, however, is the pleasure of seeing your ideas realized when projects are successfully completed, and the satisfaction of knowing that your work will remain pleasing and useful to others in the future.

HORTICULTURAL THERAPIST

Horticultural therapy is a young and still emerging profession. It is nevertheless a time-proven practice, however, since the therapeutic benefits of peaceful garden environments have been understood since ancient times. In the 19th century, Dr. Benjamin Rush, a signer of the Declaration of Independence and often called the "Father of American Psychiatry," reported that garden settings held curative effects for people with mental illness.

Rehabilitative care of hospitalized war veterans in the 1940s and 1950s greatly expanded the practice of horticultural therapy. Today, horticultural therapy is recognized as a practical and viable treatment with wide-ranging benefits for people in therapeutic, vocational, and wellness programs.

Horticultural therapy is taught and practiced throughout the world in a diverse number of settings and cultures.

Specific Work Performed

Professionals in this area are very unique. In addition to being educated in agriculture or horticulture, they are also educated in the fields of rehabilitation, counseling, and psychology.

Horticultural therapists are taught to evaluate, rehabilitate, and train clients or patients who have mental, emotional, and/or physical disabilities using standard techniques in propagating, growing, and maintaining flowers, fruits, vegetables, trees, and shrubs or in related horticultural activities. In consultation with medical teams, horticultural therapists teach a variety of horticulture techniques for vocational rehabilitation purposes or for leisure activities.

Horticultural therapists may involve a patient in a particular phase or type of horticulture. For example, they may have the patient participate in such activities as selling plants, propagating them, or maintaining plant materials on residential or commercial properties.

Most horticulturists work with people who are also trained in the area of plant materials, but horticultural therapists also work with health care professionals.

Depending on the setting in which they are employed, horticultural therapists can work with physicians, psychiatrists, counselors, social workers, and other therapists. As members of the health care team, horticultural therapists develop profiles of their clients, including an assessment of their motor skills, communication skills, and health and psychological condition. Using these data, they plan a therapy program appropriate for each individual. They also report patient or client progress, or regression, to the other members of the health care team.

The clients or patients of horticultural therapists include a wide range of individuals with mental and physical impairments. Examples of psychological or mental disabilities that might be treated by horticultural therapy are mental retardation, emotional disturbances, social maladjustments, and substance abuse.

Physical disabilities that might be treated by horticultural therapy include blindness, hearing impairment, spinal cord injury, stroke, heart attack, and cerebral palsy. The elderly and children with low self-esteem have also benefited from working with horticultural therapists. While horticultural therapists do work with individuals, many are more likely to work with their patients in groups.

Specific Skills Required

Horticultural therapists must possess a good knowledge of general horticulture; enjoy working with plants; and enjoy sharing that knowledge with others. They must also know and understand the population with whom they work. For example, if a horticultural therapist works in a nursing home, he or she must possess a good knowledge of the aging process and the physical limitations and emotional needs of the people served.

In working with people with disabilities, one must have patience, understanding, and empathy. While enjoyable and rewarding, the work of the horticultural therapist can be physically and emotionally challenging. One must combine a love of plants with a strong interest in working with people.

Work Settings

Generally, horticultural therapists work in hospitals, nursing homes, schools, vocational rehabilitation centers, drug treatment centers, juvenile detention centers, and correctional institutions. Although horticultural therapy is a relatively new professional therapeutic field, there are over 1,000 programs in operation today throughout the world. Work days are usually eight hours long and somewhat under a therapist's control. Although weekend work may be expected, a work week is rarely longer than five days. Clients of horticultural therapists usually prefer routine, and the therapist's work schedule generally reflects this.

Employment Outlook

Because of the growing body of knowledge about the positive relationship between the recovery process and rehabilitation therapies such as horticultural therapy, the outlook for employment in this area is good. While the first horticultural therapy programs were associated with vocational rehabilitation centers, increasingly horticultural therapists are being employed in settings such as drug rehabilitation facilities, nursing homes, and correctional institutions.

As the population ages, the need for more people in the field of horticultural therapy should grow. Other changes, such as the emphasis on assisting people with mental, physical, and emotional disabilities so they can work and function in mainstream society, should increase the demand for people who can teach horticulture skills to these diverse populations for both avocational and vocational purposes.

Advancement Opportunities

Horticultural therapists work as members of medically oriented teams. In this setting, the traditional career path of being promoted to supervisor or director is rarely open to them. Advanced degrees and certification qualify horticultural therapists to teach and conduct research at colleges and universities. Like art, music, and dance therapists, some horticultural therapists write scholarly and technical publications to advance in the field and to enhance their professional reputations.

Education and Other Training Requirements

A bachelor's degree in horticulture is the minimum requirement for entering this field. Undergraduate programs in horticultural therapy are primarily taught in horticulture departments in schools of agriculture that are located at land-grant institutions. Associate degrees and certificate programs are also available.

A typical horticultural therapy curriculum should not only stress agricultural and horticulture courses but should also include courses in psychology and rehabilitation. In addition, horticultural therapists should take courses in social and behavioral sciences and specific horticultural therapy techniques and practices. Degree programs in horticultural therapy require a supervised practicum or internship.

The American Horticultural Therapy Association registers horticultural therapists. Three categories of registration exist in this profession:

- *Horticultural therapist technicians* have limited education and work under the close supervision of a registered horticultural therapist.
- *Registered horticultural therapists* have advanced degrees in horticultural therapy and at least one year of paid work experience in the field.

- *Master horticultural therapists* have completed the highest level of education in horticultural therapy, have a minimum of four years paid work experience in the field, and have demonstrated extensive educational and/or professional achievement.

Unlike professionals in other therapeutic fields, horticultural therapists are not required to be licensed in any state. However, as the profession grows, the need for certification or licensure should increase. Therefore, people engaged in this field must stay current in both horticulture and rehabilitation techniques.

See Appendix B for a list of horticultural therapy programs in the United States and Canada.

Other Personal Qualifications

Horticultural therapists should enjoy working with a wide variety of people. They must be patient, well organized, reliable, and empathetic to the problems and needs of others. Physical stamina is necessary but there are no specific requirements for strength or mobility. Horticultural therapists may work under conditions of mental and emotional stress and should be able to cope with stressful situations and difficult clients.

Salary

Salaries of horticultural therapists vary according to their employment setting and educational preparation. Specific figures are difficult to find. To place horticultural therapists within the framework of other therapy professions: In 2004, median annual earnings of recreational therapists were $32,900, the average earnings for physical therapists were $60,180, and physical therapist assistants earned $37,890.

Other Rewards

Knowing that you are improving the quality of someone's life is a major secondary benefit to a career in horticultural therapy. Work settings are typically pleasant and clients generally show their appreciation to the therapists for the skills they've learned.

Education and Training Programs in Horticulture Therapy

The following list of schools and other organizations offering training in horticultural therapy is provided by the American Horticultural Therapy Association.

Bachelor's and Master's Degrees

Kansas State University
Department of Horticultural Therapy
2021 Throckmorton Hall
Manhattan, KS 66506-5506
oznet.ksu.edu/horttherapy/

Rutgers University—Cook College
Department of Plant Biology
Foran Hall, 59 Dudley Road
New Brunswick, NJ 08903
aesop.rutgers.edu

Virginia Tech
Dept of Horticulture
Blacksburg, VA 24061
hort.vt.edu/undergraduate/undergraduate.htm

Horticultural Therapy Certificate Programs (AHTA-Accredited)

Chicago Botanic Gardens
Horticultural Therapy Certificate of Merit
1000 Lake Cook Road
Glencoe, IL 60022
chicagobotanic.org

Cook College—Horticultural Therapy Certificate Program
Rutgers, the State University of New Jersey
Foran Hall, 59 Dudley Rd.
New Brunswick, NJ 08901
http://aesop.rutgers.edu/~horttherapy/

Horticultural Therapy Institute
P.O. Box 461189
Denver, CO 80246
htinstitute.org

Legacy Health System
1015 NW 22d Avenue, 5E
Portland, OR 97210
lhs.org

Minnesota Landscape Arboretum
3675 Arboretum Drive
Chaska, MN 55318-9613
umn.edu

New York Botanic Garden
200th Street and Kazimiroff Boulevard
Bronx, NY 10458
nybg.org

Providence Farm
1843 Tzouhalem Road
Duncan, BC
Canada V9L 5L6
providence.bc.ca

University of Cincinnati—Clermont College
4200 Clermont College Drive
Batavia, OH 45103
clc.uc.edu

Horticultural Therapy Courses (One Semester Class or More Offered)
Cabrillo College
Department of Horticulture
6500 Soquel Drive
Aptos, CA 95003
cabrillo.edu

Central Piedmont Community College
P.O. Box 35009
Charlotte, NC 28235
cpcc.edu

Colorado State University
Department of Horticulture & Landscape Architecture
Room 111 Shepardson
Fort Collins, CO 80523-1173
htinstitute.org

Edmonds Community College
20000 68th Avenue, West
Lynnwood, WA 98036-5999
edcc.edu

Hawkeye Community College
Attn: Horticulture Dept.
1501 E. Orange Rd., Box 8015
Waterloo, IA 50704-8015
hawkeye.cc.ia.us

Kansas State University
Division of Continuing Education
241 College Court Building
Manhattan, KS 66506-6007
dce.ksu.edu

Merritt College
12500 Campus Drive
Oakland, CA 94619
merritt.edu

Oklahoma State University—Oklahoma City
Division of Horticulture Technologies
400 North Portland
Oklahoma City, OK 73107
osuokc.edu

Murray State University
Applied Science Building
School of Agriculture
401 South Oakley
Murray, KY 42071-3345
murraystate.edu

Temple University—Ambler College
Department of Landscape Architecture & Horticulture
580 Meetinghouse Road
Ambler, PA 19002
temple.edu

Tennessee Technological University
School of Agriculture
P.O. Box 5034
Cookville, TN 38505
tntech.edu

University of Massachusetts at Amherst
Durfee Conservatory and Gardens
Dept. Plant and Soil Sciences
100 French Hall
Amherst, MA 01002
umass.edu

Wake Technical Community College
Continuing Education Registration
9101 Fayetteville Road
Raleigh, NC 27603-5696
waketech.edu

C H A P T E R

4

CAREERS IN BOTANY AND RELATED FIELDS

Plants are not only complex living organisms but also chemical factories. They produce many chemicals useful to humans, and have been described as the first step in this process that is essential to all other life. In addition to producing food, plants also generate the raw materials for paper, building materials, solvents and adhesives, fabrics, medicines (some used to treat certain types of cancer), and many other products.

In general, botanists study the chemical composition of plants and understand the complex chemical combinations and reactions involved in the metabolism, reproduction, growth, and heredity of plants. The nature of botanists' work depends to a large degree on the area in which they specialize and the setting in which they conduct it.

SPECIALIZATIONS IN BOTANY

According to the Botanical Society of America (BSA) the specializations within the field of botany can be divided into three major categories with numerous disciplines related to each specialization. The BSA specializations and their related disciplines of study include plant biology specialties, applied plant specialties, and organismic specialties.

Plant Biology Specialties

Plant biology specialties include the following:

- **Anatomy**   The study of microscopic plant structure (cells and tissues)
- **Biochemistry**   Analysis of the chemical aspects of plant life processes, including the chemical products of plants
- **Biophysics**   Application of physics to plant life processes
- **Cytology**   The study of the structure, function, and life history of plant cells
- **Ecology**   Examination of the relationships between plants and the world in which they live
- **Genetics**   The study of plant heredity and variation. Plant geneticists study genes and gene function in plants.
- **Molecular biology**   Investigation of the structure and function of biological macromolecules, including the biochemical and molecular aspects of genetics
- **Morphology**   Examination of the macroscopic plant form related to the evolution and development of leaves, roots, and stems
- **Paleobotany**   The study of the biology and evolution of fossil plants
- **Physiology**   Analysis of the functions and vital processes of plants. Photosynthesis and mineral nutrition are two examples of subjects studied by plant physiologists.
- **Systematics**   Examination of the evolutionary history and relationships among plants

Applied Plant Specialties

Applied plant specialties include the following fields:

- **Agronomy**   Application of plant and soil sciences to increase the yield of field crops
- **Biotechnology**   Use of biological organisms to produce useful products. Plant biotechnology involves inserting desirable genes into plants and having those genes expressed.
- **Breeding**   Development of better types of plants.

- **Economic botany** The study of plants with commercial importance, including harmful and beneficial plants and plant products
- **Food science and technology** Development of food from various plant products
- **Forestry** Forest management for the production of timber and conservation
- **Horticulture** The production of ornamental plants and fruit and vegetable crops. (Landscape design is an important subdiscipline of horticulture.)
- **Natural resource management** The responsible use and protection of natural resources
- **Plant pathology** The study of plant diseases, including the biological aspects of disease and disease management or control

Organismic Specialties

Organismic specialties include the following areas:

- **Bryology** The study of mosses and similar plants
- **Lichenology** Examination of the biology of lichens, dual organisms composed of both a fungus and an alga
- **Microbiology** The study of microorganisms
- **Mycology** The study of fungi
- **Phycology** The study of algae
- **Pteridology** The study of ferns and similar plants

Another method of identifying the five basic areas that concern botanists is to classify them as follows:

1. Naming and classifying plants
2. Explaining the principles underlying the diversity of plant life
3. Identifying the mechanisms that control the precise patterns of growth
4. Identifying the types of plants that grow under different conditions of temperature, soil, moisture, and nutrients in order to determine which plants will produce most abundantly
5. Studying the relationship of plants to their environment in order to determine how harmful changes in the environment can be reversed

Naturally, these issues relate to the five major career fields in botany: plant taxonomists, plant ecologists, plant morphologists, plant physiologists, and plant cytologists. The next section of this book will examine these five occupational areas in terms of the work performed, the work environment, the employment outlook, how to prepare for a career in botany, desirable personal qualifications, and salary and other benefits.

THE WORK PERFORMED BY BOTANISTS

The work of botanists varies depending on which of the five major areas of the field they specialize in. To gain a better understanding, it is best to know the nature of the work that can be performed in each specialization.

In general, botanists study the development and life processes of plants. They also study the physiology of plants (the functions and life processes of plants such as photosynthesis and nutrition), the anatomy of plants (plant cells and tissues), the morphology of plants (the evolution and development of leaves, roots, and stems), plant heredity, the environment, plant distribution, and the economic value of plants.

Overall, botanists study the behavior of plants from the chromosome level to the reproduction process. They also study plant structures that are internal (cellular) and external (leaves and stems). In addition, botanists study the mechanics of plants (the structure of plants and how they work) and the biochemistry of plants (the chemical aspects of plant life, including the chemical products of plants and/or plant cells).

Following is a brief overview of specific careers within the five major areas of botany.

PLANT TAXONOMIST

Plant taxonomists are botanists who work to identify, describe, classify, and name plant species. In so doing they provide all plant scientists with a consistent means of communication. The significance of this work is readily understood when one considers the importance of knowing the correct name, description, and classification of a plant species in almost any field within botany and horticulture. For example, in the research work of the plant physiologist, knowing the exact classification of a plant species is essential to understanding the physical characteristics of a specific plant. For the nursery grower, this knowledge is central to using the correct propagation and growing techniques, as well as knowing how to sell the plant material for proper installation in a landscape design. Likewise, the floral designer must know the classification and characteristics in order to use a particular flower or plant in an arrangement designed for certain conditions.

In addition to attempting to inventory the earth's plant resources, plant taxonomists also identify new plant species previously unknown to the field. Such discoveries and identifications may potentially lead to the identification of new food crops or new drugs which can cure or treat human diseases.

The work of plant taxonomists may also produce a new source of plant genes. The identification of new gene sources may ultimately improve the survival of endangered plant life. By using the discoveries of plant taxonomists, plant geneticists may be able to make endangered plant species stronger and more resistant to predators.

The Work Environment

Plant taxonomists must coordinate their work with the seasonal demands of the plant materials they study. For this reason, they generally aren't able to work regular hours. In some cases, they may be exposed to unsafe or unhealthy working conditions that require safety precautions.

Most plant taxonomists are engaged in field work, which involves strenuous physical activity as well as environments that are sometimes unsafe or less than comfortable. This is particularly true if their research and work is conducted in areas such as rain forests, deserts, or countries with primitive or unstable living conditions. Therefore, plant taxonomists should enjoy the outdoors and be interested in traveling to remote places.

Some of the tasks performed by plant taxonomists involve working with sophisticated equipment in laboratories and using strict scientific protocol procedures. Plant taxonomists are employed by colleges and universities as well as by organizations like botanical gardens and arboreta.

PLANT ECOLOGIST

Plant ecologists are botanists who work to understand the relationships between plants and the world in which they live. Because these interactions are complex, the plant ecologist studies the influences of such things as population size, pollutants, rainfall, temperature, and altitude on plant organisms. In some respects, the work of plant ecologists is like that of plant taxonomists. For example, plant ecologists inventory plant resources. Unlike taxonomists, however, plant ecologists categorize groupings of plant species. They are interested in learning the history of these plant communities and how their paths across the world have changed over time.

Plant ecologists examine such things as the environmental factors that regulate plant growth. They also study the responses of plants to favorable and unfavorable environmental conditions. In many cases, plant ecologists attempt to understand the natural balance between plants and their environment, while in others they attempt to manipulate this balance to further understand the relationship or achieve a better result.

One of the overriding concerns of plant ecologists is determining the most efficient methods of using plant resources. In order to do this, they draw on their comprehensive knowledge of plant taxonomy, physiology, and anatomy as well as their understanding of geology and meteorology. Plant ecologists must also rely on their strong background in chemistry.

The Work Environment

Some plant ecologists may be able to work more regular hours than plant taxonomists; however, they too must coordinate their work with the seasonal demands of the plant materials and the organizational demands of their employer or client. This is particularly true when environmental impact studies are needed in advance of, or in response to, construction projects.

In some cases, plant ecologists may be exposed to unsafe or unhealthy working conditions. When plant materials have to be studied after an environmental disaster, for example, the conditions in which the plant life is found may not be healthy or safe for the botanist. However, they take safety precautions when working under such conditions.

Most plant ecologists engage in field work, which can involve strenuous physical activity as well as more sedentary laboratory work. Some of the tasks performed involve data collection, statistical analysis, and the use of sophisticated monitoring or measurement equipment. Plant ecologists are employed by government regulatory agencies and departments as well as by colleges and universities. Some may also find work with nonprofit environmental organizations.

PLANT MORPHOLOGIST

From what you have read so far, and what you have observed in nature, you know that plant life comes in all forms, from the most microscopic organisms to the splendor of majestic trees. Botanists who have an interest in why and how these plant species differ are called plant morphologists.

Plant morphologists study the manner in which cells are arranged to give each plant its form. Their interests lie in the wide range of plant forms that exist in nature. To determine the cellular arrangement of plants, plant morphologists study how different plants transport and conduct food and water. Considering the differences between cacti and seaweed, for example, one can readily see how differently the cells meet the requirement of supplying these plants with nutrients and water.

A branch of plant morphology known as morphogenesis is the study of the reproductive structures of plants to determine the mechanisms involved in how a particular variety develops. Plant morphologists know that cells divide in precise planes and enlarge in precise locations within a developing seed at regular rates. In some species, they have learned to modify some of these steps and would like to know more about the manner in which the steps are controlled and directed within the plant itself. The true value of the work of the plant morphologist is that his or her work will provide increased knowledge about all cells.

The Work Environment

In general, plant morphologists work in research laboratories. However, they are like almost all other botanists in that they must also conduct field work. Consequently, their working hours may sometimes have to meet the demands of the plant species they are researching. For example, conducting research on the reproductive process of a particular plant form requires different conditions for pollination and germination. The very nature of this process may dictate when the researcher must be available to make observations or apply treatments.

Plant morphologists are less likely to be exposed to unsafe or unhealthy working conditions than plant taxonomists and plant ecologists. However, their research may require them to work with chemical products that necessitate strict safety procedures.

The required combination of field work and laboratory work can lead to a strenuous workload for plant morphologists. Laboratory work often involves many hours spent analyzing slides under various types of microscopes, or using special techniques for analyzing the cellular structure of the plant material. In addition, these botanists use a variety of monitoring and measurement devices to collect their data. The data collection then requires statistical analysis and report writing.

Plant morphologists may be employed by government agencies and private industry, particularly the timber industry, and many also find work in organizations such as botanical gardens and arboreta. The majority of plant morphologists, however, are employed in colleges and universities.

PLANT PHYSIOLOGIST

Plant physiologists are botanists who study the internal functions and processes of plants. The process most well known to the layperson is photosynthesis, but plant processes also include growth, respiration, circulation, excretion, movement, reproduction, and other functions. In order to study these processes, plant physiologists conduct research on the cellular structure as well as the organ-system functions of plants.

The research of plant physiologists has two major goals. The first is to develop a better understanding of the biological, chemical, and physical processes that are basic to plant life. The second is to regulate and control these activities in order to shape plant growth and development. Because basic processes govern the diversity of plant life, by understanding these processes, it is possible to learn to control when certain plants will flower. For example, by controlling the environment to regulate certain processes, growers can assure that lily plants will be in bloom for Easter or that bedding plants will flower at the appropriate time. Likewise, fruits and vegetables can be controlled to give them certain characteristics that will increase their value in the marketplace.

Plant physiologists conduct their studies of plant processes under either normal conditions for the species or abnormal conditions. This decision is determined by the nature of the study and by the effects of the internal and external environmental factors on the life processes of plants.

The Work Environment

Some plant physiologists work regular hours in a research laboratory; however, most are like other botanists and must coordinate their work with the demands of the plant materials and research requirements. Consider the wide variety of growing conditions for various species of plants. In order to conduct research on the reproduction process, certain periods of dormancy must be simulated regardless of the season or time of day. This may require the researcher to cover and uncover plant specimens at various times during the day or night—and to do so seven days a week.

Like plant morphologists, plant physiologists are less likely to be exposed to unsafe or unhealthy working conditions than some other botanists. However, their research may require them to work with chemical growth regulators or other chemical products that require strict safety procedures be followed.

Many plant physiologists are engaged in field work as well as laboratory work. Like other botanists, their work can be both strenuous and sedentary. Laboratory work often involves time spent analyzing slides under various types of microscopes or using precise staining techniques. In addition, plant physiologists use a variety of monitoring and measurement devices to collect their data. The data collection then leads to statistical analysis and report writing.

They may find work with government agencies, private industry, and organizations such as botanical gardens and arboreta. However, the majority are employed as faculty at colleges and universities.

PLANT CYTOLOGIST

Botanists who study plant cells are known as cytologists. They study the structure, function, and life history of plant cells. Cytologists select and section minute particles of plant tissue for microscopic study because they are concerned with phenomena at the cellular level. For example, if the genetic code present in the nucleus of a cell directs the synthesis of a particular protein, the cytologist wants to know the exact relationship between the steps in the photosynthesis and respiration and the molecules of the membranes associated with this process. Thus, they are concerned with the physical arrangement of DNA in a particular species of plants.

Some of the other areas cytologists study include the cells concerned with reproduction as well as the means by which chromosomes divide or unite. They also examine the formation of sperm and eggs in a variety of plant species and the physiology of unicellular organisms. These studies ascertain physical and chemical factors involved in the growth of the plant material.

Cytologists also evaluate exfoliated, aspirated, or abraded cells, or assess the hormone status and the presence of atypical or malignant changes in various plants. Because cytologists are concerned with plant tissue culture, they are able to grow entire plants from single cells in some laboratories.

The cytologist's research has the potential to expand biotechnology fields in horticulture, forestry, and plant pathology. For example, botanists working in this area might develop a blight-resistant strain of a species of tree from a single cell. This means that endangered trees which have been afflicted by disease or fungus can be propagated very quickly.

Because of the relative ease in studying plant cells, as opposed to animal cells, the research of cytologists continues to contribute to the body of knowledge necessary for advancements in fields like pharmacology.

The Work Environment

Cytologists generally work in research laboratories. Although they do not conduct field work to the same extent as plant taxonomists and plant ecologists, they coordinate some aspects of their work with the demands of the plant materials in the field.

Like plant morphologists and plant physiologists, cytologists are less likely to be exposed to unsafe or unhealthy working conditions. When their work does require them to be exposed to various chemicals, they follow strict safety procedures to assure their own safety and that of the technicians in their laboratories.

Laboratory work is often sedentary and individualistic. Because the study of cells involves considerable time analyzing slides with various types of instruments, under various types of conditions, the work must be precise. Cytologists employ a range of techniques, such as staining, to make cell structures visible or to differentiate parts of cells. In addition, monitoring and measurement devices must be calibrated to assure that the data collected during these research studies is accurate. Statistical analysis and report writing are standard aspects of the work of cytologists.

Cytologists are employed by government agencies and private industry, particularly the pharmaceutical industry. However, the majority find jobs at land-grant and research universities.

EMPLOYMENT OUTLOOK FOR BOTANISTS

Job prospects for botanists are generally good. Employment opportunities vary over time depending partly on the status of state and national economies. Some fields are more competitive than others, but challenging positions are usually available for well-trained plant scientists.

Only 10 percent of all botanists are employed in private industry. Drug companies, chemical companies, the oil industry, food companies, and lumber and paper companies are but a few of the major industrial employers of botanists. Areas of job growth include the genetic research industry, where the search for new drugs and medicines and useful genes for improving food crop plants should continue to create a demand for botanists interested in both laboratory study and in botanical exploration. Botanists who focus on environmental processes and problems should also continue to be in great demand by industry as the public's concerns about the environment and pollution increase.

Approximately 40 percent of botanists work in government agencies, and federal and state agencies should continue needing botanists in many different areas. The demand for botanists with training and experience in areas such as medicinal plant research, plant diseases, and plant genetics and breeding should be on the rise as well, particularly with the U.S. Department of Agriculture. Currently, the Department of the Interior employs botanists with interests and training in plant ecology and conservation. Even agencies such as the State Department, the Smithsonian Institution, the Environmental Protection Agency, and the National Aeronautics and Space Administration should continue to have a need for qualified botanists. In addition to the federal agencies listed above, all 50 state governments employ botanists in positions similar to those in the federal government. Slower growth than in industry can be expected due to current political and governmental priorities and possible budget cuts, but job stability will probably be good because of employment on long-term research projects.

Academic institutions—from high schools and technical training schools, to community colleges and four-year colleges and universities—should likely continue to employ the majority of botanists. More openings, however, should arise in high schools, technical schools, and community colleges for general botanists—those less interested in specialized areas of botany and in research. Most institutions at this level need people who can teach a wide range of plant-related courses, and the time and equipment used to conduct research in these schools are usually very limited.

Over 50 percent of all botanists are employed in higher-education institutions. Four-year colleges and universities should continue to need botanists with a wide variety of specializations. Almost all of these schools offer courses and degrees in botany or plant science; thus, faculty positions should be available for botanists with different specialties. Research and administrative positions are also common in these educational institutions. Just as in government positions, slower growth may be expected in academia than in industry due to political and social trends and possible budget cuts that affect the availability of funds for both teaching and research positions in education. Job stability, however, should remain good due to long-term projects and the continuing need for qualified faculty.

Many other careers for botanists do not involve teaching or research. Some botanists work in administration and marketing for biological supply houses, seed companies, biotechnology firms, pharmaceutical manufacturers, and scientific publishers. Some work as scientific writers, illustrators, and photographers.

Any economic recession would tend to limit opportunities for botanists, thus the number of jobs depends on local, state, and national economies. Some specializations may be in demand more than others, and any oversupply of persons with training in a particular field should make competition for positions greater. However, the field of botany is so diverse that challenging positions will likely be available for any well-trained botanist.

PREPARING FOR A CAREER IN BOTANY

Now that you're familiar with the various specialties that exist as career options within the field of botany, it's time to consider how best to prepare for a career in one of these areas.

High School Preparation

Students in high school should take a college preparatory curriculum, including courses in English, mathematics, a foreign language, physics, chemistry, and biology. Courses in social science and the humanities should also prove valuable for later college-level studies. Any classes that can help you develop good communication skills are also recommended.

If you're planning to pursue a bachelor's degree in botany or horticulture, your high school curriculum should include the following courses:

- Algebra I and II
- Trigonometry
- Biology
- Physics
- Social studies (3 units)
- Fine arts or humanities (1 or 2 units)
- Computer programming or computer applications
- Geometry
- Calculus
- Chemistry
- English (4 units)
- Foreign languages (2 or 3 units)

Other courses that may be helpful include economics, history, and public speaking. It's also recommended that prospective botany and horticulture students take advanced placement or honors level courses, and that you set a goal of achieving combined scores of at least 1000 on the SAT exam or 20 on the ACT exam. You should, of course, be proficient in computer use.

Admissions officers at colleges and universities also look for well-rounded students. Extracurricular activities during high school can reflect this. Being a member of, and holding an office in, science clubs demonstrates strong and consistent interests. However, participation in athletics, service organizations, and cultural activities are also important.

Extracurricular activities that may prove valuable include participation in science clubs and science fairs. Summer jobs and internships in any area related to biology can also help. This might include working in parks, laboratories, public and private gardens, plant nurseries and greenhouses, farms, and camps. Hobbies such as gardening, photography, and camping are also useful.

Prior to applying to any college or university, you should gather as much information as possible from each school you're considering. Your counselor and library can also help with this. Many schools do not have separate botany departments but instead teach botany within a department of biology. Check out the web sites of schools that interest you, and visit the campuses of as many as possible. Ask to meet with some of the botany faculty to discuss your career goals and options and how they feel their department can help you fulfill those goals. The Botanical Society of America also provides information about education and career opportunities. (The society's contact information is listed in Appendix B at the end of this book.)

College Curriculum

Over 90 colleges and universities in the United States and Canada offer undergraduate degrees in botany. The specific courses you take in college will depend on the school's curriculum and your own interests. Most colleges and universities require a core program in biology before you can enroll in any specialized botany courses. At other colleges, however, you may be able to take botany courses immediately.

In general, the first two years concentrate on mathematics and the physical sciences, and involve introductory biology and chemistry classes as well as courses in English and the social sciences. The last two years include required courses in the major. In addition, students take required electives (which help focus the undergraduate degree in your major) as well as "free" electives. "Free" electives are courses that the student wishes to take. These courses can either be any class outside of your major department or those on a list of electives approved by your department.

To be best prepared for the job market or for graduate school, you should take courses in writing and literature, the arts, humanities, and social sciences in addition to your general biology and specialized botany courses. Most curricula also require that you take courses in statistics, chemistry, and physics, as well as general math classes up through calculus. You should also be proficient in using a computer and a wide variety of software programs. In addition, some schools may require fluency in a foreign language, which could be especially important for any field botanist interested in working outside the United States.

Summer jobs, internships, and cooperative education can provide important work experience for students. Positions exist in government, college and university laboratories, agricultural and biological research stations, and in private industry.

If at all possible, arrange to do an undergraduate research project under one of your professors. This might be an independent research project of your own or one that assists a faculty member with his or her research. This type of activity can help you choose which areas of botany you like best or least, and give you invaluable experience and insight into how science works. Obviously, it can also aid you in deciding whether graduate school and a future as a research botanist is the best choice for you.

Other Personal Qualifications

Botanists must have clear and concise verbal and written communication skills, a high level of curiosity about the world around them, be creative in solving problems, and yet have a good understanding of the scientific method and the rigors of scientific research. Generally, they should be detail oriented, and careful and precise in their work. Some specialties in botany require physical strength and stamina, while others may be sedentary. All areas require mental stamina due to long hours in the laboratory and in interpreting data. Good interpersonal skills, given the amount of collaborative work and interaction with colleagues and students, is a bonus. Some specialties, however, may be better suited to the person who prefers to work alone with only limited interaction with others.

Botanists are often required to understand and operate a variety of scientific equipment ranging from the simple to the sophisticated. Computer skills are essential in the scientific world. A clear understanding of a variety of software programs is a must, and some skills in programming may prove beneficial. The ability to work with microscopes is important as well, considering that botanists work with a range of instruments, from the simple microscopes found in most high school biology laboratories to sophisticated electron microscopes that are common in most university and industry laboratories today.

Other beneficial skills might include staining techniques, radioisotope analysis, digital imaging analysis, satellite imaging, telemetry, and cell and tissue culture. Each of these procedures requires specialized equipment, and the ability to operate that equipment properly is essential. These skills may be taught in specific courses in college or they may be learned by working in a research laboratory with technicians familiar with the equipment.

Botany offers numerous interesting and fulfilling career opportunities, with work that is often as varied as the botanist's surroundings. Because of this great diversity, many people with different backgrounds, capabilities, and interests find careers in botany rewarding.

MATCHING YOUR PERSONAL INTERESTS WITH OCCUPATIONAL AREAS IN BOTANY

Depending on your personal interests, there is a variety of botany specializations for which you are probably suited. The following are some of the specializations in which botanists work and the type of people who find these specializations interesting and satisfying.

For People Who Enjoy Research and the Outdoors

Because of the nature of their work, botanists in the following categories generally do not work, day in and day out, in the confines of research laboratories. Most work outside to identify new plant forms in areas such as tropical rain forests, sea coasts and oceans, rivers and river beds, and freshwater bodies. Therefore, the following are occupational areas within the botany career field for people who enjoy the outdoors and who enjoy travel to remote areas of the world.

- Plant taxonomists identify, name, and classify plants. (See the "Plant Taxonomist" section earlier in this chapter.)
- Plant ecologists study the relationships between plants and the environment in which they live. (See the "Plant Ecologist" section earlier in this chapter.)
- Aquatic botanists identify, describe, classify, and name plant life in either salt water or fresh water. (See the "Aquatic Botanist" section later in this chapter for more information.)
- Paleobotanists study the evolution of fossil plants.

For People Who Enjoy Chemistry

The following occupational areas are well suited for people who not only enjoy the study of plants but also enjoy observing, analyzing, and manipulating chemical reactions.

- Plant physiologists study the functions and vital processes of plants. Some of these processes include photosynthesis and mineral nutrition. (See the "Plant Physiologist" section earlier in this chapter.)
- Plant biochemists study chemical aspects of plant life processes, including the chemical products of plants (phytochemistry).
- Molecular biologists study the structure and function of biological macromolecules. Molecular biology includes the biochemical and molecular aspects of genetics.
- Chemotaxonomists, in a subdiscipline of plant taxonomy, use the chemicals produced by plant groups to aid in their classification.

For People Who Enjoy Mathematics

For those who enjoy both mathematics and plants, there are areas of botany where numerical modeling of biological systems and the application of physics principles contribute to the increasing body of knowledge about plant life, as shown in the following:

- Biophysicists specialize in applying the principles of physics to plant life processes.
- Plant geneticists study plant heredity and variation in terms of genes and gene function in plants. (See the "Plant Geneticist" section later in this chapter.)
- Systems ecologists work with the numerical aspects of plant life and their relationship to the environment. They employ mathematical models to demonstrate concepts such as nutrient cycling in plant species.

For People Concerned about the World Food Supply

People concerned about the world food supply may be interested in the following areas:

- Plant pathologists study diseases of plants. They are concerned with both the biological aspects of disease and with disease management or control. (See the "Plant Pathologist" section later in this chapter.)
- Plant breeders seek to develop better types of plants. Breeding involves selecting and crossing plants with desirable traits such as disease resistance.

For People Who Enjoy Plant Structures at the Molecular Level

People who find complexity of form and design interesting may enjoy occupations that deal with the molecular level of botany. Some of these occupations include:

- **Bryologists**   Botanists in this area study all aspects of mosses and similar plants, and are involved in the identification, classification, and ecology of these plant species.
- **Cytologists**   Botanists in this area study plant cells and their structure, in addition to their function and life history.
- **Microbiologists**   Microbiologists study microorganisms such as bacteria. They sometimes specialize by organisms or by branch of biology.
- **Mycologists**   Botanists in this area study the biology of fungi. Fungi have a tremendous impact on the environment because they are crucial in recycling dead organic material. Some fungi are important producers of biological products such as vitamins and antibiotics.
- **Phycologists**   Phycologists study algae. As noted previously, algae are the base of the food chain in the aquatic environments of the world. Those phycologists who study algae in oceans are sometimes called marine botanists.
- **Pteridologists**   Botanists in this area study ferns and similar plants.

For People Interested in Artistic Uses of Plants

People who find the artistic use of plant forms and color interesting may enjoy a number of occupational fields in ornamental horticulture. Some of these fields include landscape design, interior landscape design, and floral design. They are discussed in detail later in this book.

SALARIES

Salaries for botanists depend on their level of education and area of specialization, the geographic area of employment, and whether the botanist is employed in the private sector or by a government agency.

The Botanical Society of America reports that in 2003 the median income for botanists with less than one year of experience was $33,000; for those with 30 years or more, the median salary was $108,000. The median for all nonsupervisory positions was $48,000. For those in supervisory positions, the median income was $126,500.

Examples of average salaries for specific careers in botany are as follows: intermediate research technician, $33,000; intermediate researcher, $50,250; laboratory manager, $53,000; laboratory director, $90,000; research section head, $108,387. Government section heads earned an average of $98,000. Research managers earn an average of $139,000, while research vice presidents/directors earn $142,000.

Botanists employed in academia earn salaries based on their rank and the type of institution at which they teach. In general, professors earn an average of $118,000; distinguished researchers, $126,000; and college department heads, $129,000.

Other Rewards

There are benefits other than salary associated with a career in botany. Many positions in botany provide rewards such as individual freedom, intellectually stimulating work associates, a pleasant work environment, varied work assignments, and an opportunity to travel. The great sense of accomplishment that comes from working in a scientific field that can bring great benefits to the world and its inhabitants is probably one of the most satisfying aspects of pursuing a career in botany.

CAREER FIELDS RELATED TO BOTANY AND HORTICULTURE

The career fields of botany and horticulture are so broad and diverse that there are specific career options that cannot be easily classified under a particular heading. Though not specific to either career field, the following divisions of botany and horticulture should also be considered.

PLANT PATHOLOGIST

Plant pathologists are botanists who study diseases that afflict plants. Plant pathologists conduct studies on the nature, cause, and control of plant diseases as well as the decay of plant products. Often, their studies compare healthy plants with diseased specimens to determine the agent responsible for the affliction. They are concerned with the spread and intensity of disease under different conditions of soil, climate, and geography, and frequently predict outbreaks of plant disease and try to determine which kinds of plants and insects will transmit it. Plant pathologists are concerned with both the biological aspects of diseases and with disease management and control.

Plant pathologists attempt to isolate disease-causing agents and study the habits and life cycles of various plants. Many times, their goal is to destroy or control the agent that causes the disease. For example, many plant pathologists have studied Dutch elm disease in an attempt to identify the agent and control its spread. Unfortunately, many elm trees had to be destroyed in order to halt the spread of the disease. Plant pathologists also test possible disease control measures under experimental conditions in their laboratories and research fields. They study the comparative effectiveness of different treatments and make recommendations based on practicality, economy, and safety.

PLANT GENETICIST

Plant geneticists are botanists who study plant heredity and plant variation. They perform experiments to identify the biological laws and mechanisms, as well as the environmental factors that are central to the origin, transmission, and development of inherited traits in plants.

In addition, plant geneticists analyze the determinants of specific inherited traits. These traits may include such factors as color differences, plant size, and disease resistance. Knowledge gained through this research can be used to improve or understand the relationships of heredity, maturity, fertility, or other factors.

Botanists who work in the area of plant genetics devise methods for altering or producing new traits in plant species. They make use of chemicals, temperature, light, or other means to study the plant/environment relationship.

Plant geneticists may specialize in particular branches of genetics such as molecular or population genetics. Probably the best-known plant geneticist in the area of molecular biology is Barbara McClintock, who won the Nobel Prize in 1983 for her pioneering work in corn genetics. As a plant molecular biologist, Dr. McClintock's work involved the transposable elements of genes (referred to as "jumping genes") and the plant/environment interaction. These discoveries led to the development of a stronger species that can resist disease, pests, and climate changes, which has contributed to the development of a more durable food crop for distribution and consumption. The techniques and principles developed by Dr. McClintock continue to be applied to other species and types of studies by plant scientists.

Approximately ten years ago, the first genetically altered food crop reached store shelves. A tomato that has a much longer shelf life than other tomatoes without loss of flavor signaled the beginning of a new career field for botanists. Through plant biotechnology, desirable genes are modified or inserted into plants and are expressed in the new crop. Biotechnology uses this genetic manipulation of biological organisms to reproduce useful products.

PLANT BREEDER

The work of plant breeders is related to that of plant geneticists. In plant breeding, the research of plant geneticists is used to develop and improve varieties of crops. Plant breeders improve specific characteristics, such as yield, size, quality, maturity, and resistance to frost, drought, disease, and insect pests. They accomplish this by utilizing both the principles of plant genetics and the knowledge of plant growth.

Plant breeders develop varieties and select the most desirable plants for continued reproduction using methods such as inbreeding, crossbreeding, backcrossing, outcrossing, mutation, or interspecific hybridization and selection. They select offspring of plants that have the desired characteristics and they continue the breeding and selection process until the desired characteristics are consistently expressed by the majority of new plants.

Through their work, plant geneticists and plant breeders have advanced the field of biotechnology in the area of botany. Although biotechnology is viewed primarily as the genetic modification of living organisms to produce useful products, plant biotechnology involves inserting desirable genes into plants and having those genes expressed. As a result, the research of plant geneticists has the potential to improve domestic plants.

AQUATIC BOTANIST

Aquatic botanists study plants that live in water. Aquatic botanists who identify, describe, classify and name plant life in fresh water are also known as limnologists. Those who study salt-water organisms are known as phycologists because they study algae. Their work is of significance because algae are near the base of the food chain in the aquatic environments of the world. Botanists who study salt-water plant life are also called marine botanists.

AGRICULTURAL ENGINEER

Agricultural engineering can be described as one of the basic engineering disciplines with the closest relationship to horticulture and botany. This is so because agricultural engineers ensure tomorrow's food production.

They work with horticulturists to design equipment and processes that plant, harvest, produce, and distribute food stuffs. As agricultural engineers develop new tools, it becomes easier and more practical to produce, process, and distribute food. Using scientific principles, agricultural engineers design systems and equipment to manage the various resources that horticulturists and botanists need to produce food. These resources include soil, water, air energy, and engineering materials. Agricultural engineers apply their skills across the entire food production chain, from the protection of natural resources to the preservation of food products.

Some agricultural engineers are employed in the increasingly technological agricultural industry. Others run experimental farm stations and research laboratories, usually at land-grant universities, which benefit the agricultural community and industry. Still others work as consultants for projects in foreign countries.

With society's increasing concern about world hunger, the agricultural engineer plays an important role on the team of botanists and horticulturists who plan and implement new technologies for food production.

Unlike the preparation of horticulturists and botanists, agricultural engineers concentrate their studies in mechanical, civil, electrical, chemical, and other types of engineering, in addition to biology and horticulture training. Universities that specialize in preparing agricultural engineers are located at most land-grant universities, where courses in both engineering and agriculture are offered.

C H A P T E R

5

TEACHING OPPORTUNITIES FOR BOTANISTS AND HORTICULTURISTS

Those who pursue undergraduate degrees in botany and horticulture have the career option of teaching young people about their discipline in private and public schools, primarily at the secondary level. To teach at this level, you must not only complete an undergraduate degree in botany or horticulture, you must also become certified in the state in which you want to teach.

Likewise, people who pursue a master's degree in botany and horticulture will find that junior and community college teaching may be a viable career option. Teaching at this level involves preparing students to go directly into industry or to pursue a bachelor of science degree at a four-year college or university. This means that people who teach at this level must have a good knowledge of industry, as well as higher-education requirements beyond the associate degree level.

Those who pursue a Ph.D. degree and who are involved in research in botany and horticulture will find teaching at four-year universities or two-year colleges a viable career option. In addition to preparing future generations of botanists and horticulturists, members of college faculties enjoy learning and actively contribute to the body of knowledge in their field.

TEACHING AT THE SECONDARY LEVEL

Elementary teachers introduce young students to very basic facts about plants, flowers, and trees through science courses and special projects. However, it is the secondary teachers in biology and vocational education courses who introduce high school students to more in-depth knowledge about plant material. They cover such concepts as photosynthesis, propagation, and plant nutrition.

Teachers at this level can expect to supervise laboratories as well as teach classes in their subject area. Vocational education teachers, who teach horticultural skills, may often supervise student work experience in nurseries, greenhouses, and commercial businesses such as florists and landscape design firms.

In addition to classroom responsibilities, secondary teachers, regardless of the subject matter they teach, are responsible for advising students and meeting with parents. Many also advise student organizations and participate in other aspects of school life, such as serving on school and PTA committees.

The reality of a teacher's work schedule exceeds the hours of a typical school day. Course preparation, student evaluation, and school service make teaching at least a 40-hour-a-week job, in which more hours are often spent on these pursuits than on actual teaching. While there are some year-round school districts, the majority of teachers work ten months a year and have a two-month, nonpaid vacation during the summer. Many teachers take college courses during this time in order to stay current in their fields. Others teach in summer school.

In most states, teachers serve a probationary period, usually three years. After this probationary period, they are considered to be "tenured," meaning they cannot be fired without just cause and due process.

Education and Training

All 50 states and the District of Columbia require public school teachers to be licensed. Licensure is not required for teachers in private schools in most states, however. Usually, licensure is granted by the state's Board of Education or a licensure advisory committee. Secondary school teachers are licensed to teach a specific subject area to students in grades 7 through 12.

Although requirements for regular licenses vary by state, all require general education teachers to have a bachelor's degree and to have completed an approved teacher training program with a prescribed number of subject and education credits, as well as to have performed supervised practice teaching. Some states also require technology training and the attainment of a minimum grade point average. A number of states expect that teachers obtain a master's degree in education within a specified period after they begin teaching.

Almost all states require applicants for a teacher's license to be tested for competency in basic skills, such as reading and writing, not to mention teaching. Most also require the teacher to exhibit proficiency in his or her subject. Many school systems are presently moving toward implementing performance-based systems for licensure, which usually require a teacher to demonstrate a satisfactory teaching performance over an extended period in order to obtain a provisional license, in addition to passing an examination in their subject. Most states require continuing education for renewal of the teacher's license, and many have reciprocity agreements that make it easier for teachers licensed in one state to become licensed in another.

A number of states also offer alternative licensure programs for teachers who have a bachelor's degree in the subject they will teach, but who lack the necessary education courses required for a regular license. Many of these programs are designed to ease shortages of teachers of certain subjects, such as mathematics and science. Other programs provide teachers for urban and rural schools that have difficulty filling positions with teachers from traditional licensure programs. Alternative licensure programs are intended to attract people into teaching who do not fulfill traditional licensing standards, including recent college graduates who did not complete education programs and those changing from another career to teaching.

In some programs, individuals begin teaching quickly under provisional licensure. After working under the close supervision of experienced educators for one or two years while taking education courses outside school hours, they receive regular licensure if they have progressed satisfactorily. In other programs, college graduates who do not meet licensure requirements take only those courses that they lack and then become licensed. This approach may take one or two semesters of full-time study. States may issue emergency licenses to individuals who do not meet the requirements for a regular license when schools cannot attract enough qualified teachers to fill positions. Teachers who need to be licensed may enter programs that grant a master's degree in education, as well as a license.

Private schools are generally exempt from meeting state licensing standards. For secondary school teacher jobs, they prefer candidates who have a bachelor's degree in the subject they intend to teach. They seek candidates among recent college graduates as well as from those who have established careers in other fields. Private schools associated with religious institutions also desire candidates who share the values that are important to the institution.

In some cases, teachers may attain professional certification in order to demonstrate competency beyond that required for a license. The National Board for Professional Teaching Standards offers a voluntary national certification. To become nationally accredited, experienced teachers must prove their aptitude by compiling a portfolio showing their work in the classroom and by passing a written assessment and evaluation of their teaching knowledge. Currently, teachers may become certified in a variety of areas, on the basis of the age of the students and, in some cases, the subject taught. For example, teachers may obtain a certificate for teaching English language arts to early adolescents (aged 11 to 15), or they may become certified as early childhood generalists. All states recognize national certification, and many states and school districts provide special benefits to teachers holding such certification. Benefits typically include higher salaries and reimbursement for continuing education and certification fees. In addition, many states allow nationally certified teachers to carry a license from one state to another.

The National Council for Accreditation of Teacher Education currently accredits teacher education programs across the United States. Graduation from an accredited program is not necessary to become a teacher, but it does make it easier to fulfill licensure requirements. Generally, four-year colleges require students to wait until their sophomore year before applying for admission to teacher education programs. Traditional education programs for kindergarten and elementary school teachers include courses designed specifically for those preparing to teach. These courses cover mathematics, physical science, social science, music, art, and literature, as well as prescribed professional education courses, such as philosophy of education, psychology of learning, and teaching methods. Aspiring secondary school teachers most often major in the subject they plan to teach while also taking a program of study in teacher preparation. Teacher education programs are now required to include classes in the use of computers and other technologies in order to maintain their accreditation. Most programs require students to perform a student-teaching internship as well.

Many states now offer professional development schools, which are partnerships between universities and elementary or secondary schools. Students enter these one-year programs after completion of their bachelor's degree. Professional development schools merge theory with practice and allow the student to experience a year of teaching firsthand, under professional guidance.

Other Personal Qualifications

In addition to being knowledgeable in their subject, teachers must have the ability to communicate, inspire trust and confidence, and motivate students, as well as understand the students' educational and emotional needs. Teachers must be able to recognize and respond to individual and cultural differences in students and employ different teaching methods that will result in higher student achievement. They should be organized, dependable, patient, and creative. Teachers also must be able to work cooperatively and communicate effectively with other teachers, support staff, parents, and members of the community.

Earnings

The following table shows the average starting and annual salaries of public school teachers in each state for the 2005–06 academic year.

State	Beginning Salary	Average Salary
Alabama	$30,973	$38,186
Alaska	$40,027	$54,424
Arizona	$28,236	$42,905
Arkansas	$26,129	$40,495
California	$35,135	$57,876
Colorado	$31,296	$43,949
Connecticut	$34,462	$58,688
Delaware	$34,566	$50,869
District of Columbia	$38,566	$58,456
Florida	$30,696	$41,587
Georgia	$35,116	$46,526
Hawaii	$37,615	$46,149
Idaho	$25,908	$42,122
Illinois	$35,114	$55,629
Indiana	$29,784	$46,526

Iowa	$26,967	$39,284
Kansas	$28,530	$39,175
Kentucky	$28,416	$40,522
Louisiana	$29,655	$38,880
Maine	$25,901	$39,610
Maryland	$33,760	$52,331
Massachusetts	$34,041	$54,325
Michigan	$34,377	$56,973
Minnesota	$30,772	$46,906
Mississippi	$28,106	$36,590
Missouri	$28,938	$38,971
Montana	$24,032	$38,485
Nebraska	$28,527	$39,456
Nevada	$27,942	$43,394
New Hampshire	$27,367	$43,941
New Jersey	$37,061	$56,682
New Mexico	$31,920	$39,391
New York	$36,400	$56,200
North Carolina	$27,572	$43,348
North Dakota	$24,108	$36,449
Ohio	$28,692	$48,692
Oklahoma	$29,473	$37,879
Oregon	$33,396	$48,330
Pennsylvania	$34,140	$53,258
Rhode Island	$32,902	$53,473
South Carolina	$27,883	$42,207
South Dakota	$25,504	$34,040
Tennessee	$30,449	$42,072
Texas	$32,741	$41,009
Utah	$26,130	$39,965
Vermont	$25,819	$44,535
Virginia	$32,437	$44,763
Washington	$30,159	$45,724
West Virginia	$26,692	$38,360
Wisconsin	$28,800	$44,299
Wyoming	$28,900	$40,392

Estimate Source: National Education Association (2004–05).

Career Outlook

There is an increasing demand for teachers with a scientific background. Therefore, people with undergraduate degrees in botany, or biology with an emphasis in botany, should find job prospects to be very good, provided they have the necessary certification.

Not every secondary school offers vocational education programs in horticulture, thus limiting the opportunity for people with horticulture degrees to teach their specific discipline. However, if they have completed a sufficient number of biology or chemistry courses in their undergraduate preparation, they may be able to become certified to teach in these disciplines within the state system where they seek certification.

However, people with undergraduate degrees in horticulture are not limited to teaching only within the public school setting. Opportunities exist to teach horticulture skills and knowledge in nontraditional settings such as vocational rehabilitation centers or correctional institutions, making the options to teach horticulture varied.

Shortages of qualified teachers will likely continue, resulting in competition among some localities, with schools luring teachers from other states and districts through bonuses and higher pay.

Through 2014, overall student enrollments, a key factor in the demand for teachers, are expected to rise more slowly than in the past as children of the baby boom generation leave the school system. This should cause employment to rise up to 17 percent for teachers through the secondary grades. Projected enrollments should vary by region. Fast-growing states in the west— particularly California, Idaho, Hawaii, Alaska, Utah, and New Mexico— should experience the largest enrollment increases. Enrollments in the South should increase at a more modest rate than in recent years, while those in the Northeast and Midwest are expected to remain relatively steady or decline. Teachers who are geographically mobile and who obtain licensure in more than one subject should have a distinct advantage in finding a job.

Advancement Possibilities

Secondary teachers can advance through a system of promotions to become senior or mentor teachers, with higher pay and additional responsibilities. These teachers guide and assist less experienced teachers while keeping most of their own teaching responsibilities.

Some teachers also move into counseling or administrative positions. Administrative positions within a school can include department chair, assistant or vice principal, and principal. Administrative positions within the school system may include supervisor of curriculum or instruction at the local or state level.

More Information on Teaching at the Secondary Level

To learn more about teaching opportunities and certification requirements in your area of interest, contact the professional associations related to the specific discipline you plan to teach. You should also get in touch with the local school system and the state department of education in that area. Other resources on teaching careers include

National Education Association
1201 16th Street NW
Washington, DC 20036
nea.org

American Federation of Teachers
555 New Jersey Avenue, NW
Washington, DC 20001
aft.org

In addition, Recruiting New Teachers, Inc. offers guidance on how to become a teacher and provides an online job bank of available teaching positions.

Recruiting New Teachers, Inc. (RNT)
385 Concord Avenue, Suite 103
Belmont, MA 02478
recruitingteachers.org

TEACHING AT THE COLLEGE AND UNIVERSITY LEVEL

Botany faculty teach at a wide variety of colleges and universities. Some who teach at four-year institutions work at major research institutions, while others teach at land-grant colleges and universities. Still others teach at liberal arts colleges and universities. In contrast, horticulture faculty primarily teach at land-grant colleges and universities as well as two-year and community colleges.

Some faculty members pursue research as the major emphasis of their work. However, while research funding and publications may be of paramount importance in nearly every college and university, poor teaching is viewed negatively even at institutions that place primary emphasis on research.

Because the preparation of future botanists and horticulturists is vital to the profession and to the nation, teaching is at the center of any academic career. However, most four-year colleges and universities tend to emphasize the importance of faculty research, while two-year and community colleges place their primary emphasis on teaching. In the area of horticulture, however, both two-year and four-year institutions put importance on the relationship between the faculty and farmers and/or the agricultural industry.

Faculty employed at land-grant universities can have "split appointments." This means that they can have a primary appointment in the areas of teaching, research, or extension. If their appointment is chiefly in the area of teaching, they will be evaluated on their teaching innovation and abilities. Those who have a primary appointment in research will be evaluated on the quality and significance of their basic or applied research in their field. In the area of extension, faculty members are evaluated on their service to industry and the public.

An important aspect of an academic career for all faculty is interaction with colleagues and students, whether undergraduate or graduate. Faculty members enjoy a special fellowship and camaraderie among themselves and their students. This interaction usually transcends the campus to include faculty, students, and professionals around the country and the world. Through these scholarly interactions, college teachers have almost always had a leading role in the pursuit of scientific and technological advances in botany and horticulture.

Working hand-in-hand with the government and industry, faculty research has advanced the knowledge base in botany and horticulture. The rapidly increasing interest in the information and skills essential to grow and market horticultural crops is related to the world's concern for environmental quality and health. Therefore, the role of botany or horticulture faculty members has become even more significant.

Faculty are at the center of the transfer of scientific knowledge for industrial and consumer usage. Through scholarly research, innovative teaching, and major contributions to the literature of their field, faculty are instrumental in impacting the overall quality of life. Therefore, if you enjoy the process of learning and growing intellectually, if you like research and laboratory work, an academic career in the area of botany or horticulture may be a very appropriate career path.

Where to Begin

Interestingly, an academic career does not begin with your first faculty appointment. It actually begins when you start to think about going to graduate school. In some instances, you might be advised to obtain experience in the field before pursuing a graduate degree. This advice is not contradictory; rather, it is quite sound in today's academic environment. Experience as an undergraduate intern, as a co-op student, or as a full-time employee will provide an excellent foundation for an academic career. There are two major benefits to this approach.

First, as a future faculty member your experience will provide a better understanding of the issues that your students will encounter in the field. This is extremely important because most students will pursue career paths in business, industry, or government. Your understanding of the issues and constraints they will face will prove important in how you advise and teach them. Second, practical experience will help you focus more clearly on the objectives of your graduate study. It may even help you determine the subspecialties you wish to pursue and the best faculty person with whom to study in that area.

Upon entering graduate school, your appointment as a graduate research assistant or a graduate teaching assistant can form the basis for a successful academic career. Prestigious fellowship awards to support your graduate education may also help you gain a faculty position after you complete your graduate studies.

Today, it is almost a necessity to have postdoctoral research experience before receiving an appointment on a college or university faculty. A track record of successful research and scholarly publications is an important credential when competing for faculty positions. Your postdoctoral time, usually at a university different from the one awarding your Ph.D., provides you with additional time to build your scholarly reputation in the field.

Appointment as an assistant professor in a "tenure track" position is the point at which you become a member of the faculty. Tenure-track faculty are usually hired as instructors or assistant professors, and must serve a period under term contracts, generally for seven years. At the end of this period, tenure is granted based on a favorable review of their teaching, research, and overall contribution to the institution. Those denied tenure usually must leave the institution.

Tenured professors cannot be fired without just cause and due process. Tenure protects the faculty's academic freedom, and gives both faculty and institutions the stability needed for effective research and teaching. It also provides financial security for faculty. Some institutions have adopted post-tenure review policies to encourage the ongoing evaluation of tenured faculty.

In general, associate professors remain active in their teaching, research, and/or service to the profession. By doing so, they may qualify for promotion to full professor, which is the highest academic appointment one can receive.

Working Conditions

The working conditions in colleges and universities differ widely, since each campus has its own unique personality and style. An important determinant of these differences is the overall mission of the institution and the focus of the department. The differences also are attributable to the individuals who make up the faculty.

Every college and university catalog clearly states the institution's mission. Prospective faculty members should read those statements carefully. In addition, if the department has its own mission statement, that should be examined as well. These statements of purpose will tell you what activities the institution values, and consequently which activities it will reward.

Some institutions emphasize faculty research and publications. Faculty in these settings are expected to set a research agenda for themselves, secure outside funding from industry, foundations, and/or government agencies, publish their results, sponsor colloquia and symposiums, and become recognized experts in their specific area of engineering. Schools that are ranked among the top 25 graduate programs in the country generally fall into this category.

Other schools emphasize teaching. In these unique settings, innovative teaching techniques and the publication of textbooks are expected. The faculty/student ratio is an important consideration because the academic development of the students is critical to the mission and goals of this type of institution. This is not to say that teaching is undervalued in research-oriented universities. All institutions of higher learning strive to provide a high level of teaching. However, at institutions that focus exclusively on teaching, faculty members are held more accountable for their expertise in the classroom, as opposed to their research expertise.

In addition to teaching and research-oriented institutions, there are also institutions whose mission it is to provide extension services to farmers, industry, and home owners. Land-grant universities are excellent examples of institutions that are committed to professional practice in the various specialties of the agricultural industry. These institutions provide a valuable service to business, industry, and government.

Excellent teaching, research productivity, and extension service do not automatically mean that faculty can continue to teach and conduct research indefinitely. Most faculty have to be reviewed by a rigorous tenure process at the end of their fifth year. While some institutions either do not have tenure or have abolished it, the majority still maintain this process of peer review to assure the quality of the faculty.

Full-time college teachers usually have flexible schedules. They are usually required to teach 12 to 16 hours each week, and to attend faculty and committee meetings. Most also hold regularly scheduled office hours of three to six hours per week for student consultations. Aside from these obligations, faculty members are free to determine their own schedules for course preparation, grading, study, research, and other activities.

Some college teachers teach night and weekend classes, particularly those at two-year community colleges or institutions with large enrollments of older students. At most institutions, teachers are required to work nine months of the year. Many use the three months off to teach additional courses, conduct research, travel, or pursue other interests. Most colleges and universities have funds to support the research or other professional development needs of full-time faculty, including travel to conferences and research sites.

University faculty may experience a conflict between their responsibilities to teach students and the pressure to conduct research and publish their findings. This may be a particular problem for young faculty seeking advancement in four-year research universities. Also, recent cutbacks in support workers and the hiring of more part-time faculty have put a greater administrative burden on full-time faculty. Requirements to teach online classes also have added greatly to the workloads of postsecondary teachers. Many find that developing the courses to put online, plus learning how to operate the technology and answering large amounts of e-mail, is very time-consuming.

In 2004, about three out of ten college and university faculty worked part-time. Known as "adjunct faculty," some of these teachers have primary jobs outside of academia, in government, private industry, or nonprofit research. Others prefer to work part-time hours or seek full-time jobs but are unable to obtain them due to intense competition for available openings, and still others are employed part-time by more than one institution. Some adjunct faculty, however, are not qualified for tenure-track positions because they lack a doctoral degree.

Earnings

It is important to note that faculty salaries at both two-year and four-year institutions reflect nine- or ten-month contracts, not twelve months of income. In order to assure a year-round salary, faculty must budget in order to live on their base salary on a twelve-month basis, or they must secure summer teaching and/or research funding to increase their base salary. However, at some institutions faculty may only cover a portion of their summer salary using these means.

As faculty develop national, state, and local reputations in their field, they can augment their salaries with consulting fees. Business, industry, and government seek the expertise of college and university faculty who have established their expertise in a specific area. Royalties from books, software, and/or patents also contribute to a faculty member's overall income.

Earnings for college faculty vary according to rank and type of institution, geographic area, and field. According to a 2004–05 survey by the American Association of University Professors, salaries for full-time faculty averaged $68,505. By rank, the average was $91,548 for professors, $65,113 for associate professors, $54,571 for assistant professors, $39,899 for instructors, and $45,647 for lecturers. Faculty in four-year institutions earn higher salaries, on average, than those in two-year schools. In 2004–05, faculty salaries averaged $79,342 in private independent institutions, $66,851 in public institutions, and $61,103 in religiously affiliated private colleges and universities.

In comparing faculty base salaries in botany and horticulture, *The Chronicle of Higher Education* annually publishes an extensive breakdown of all faculty salaries by rank, discipline, and institution. This is a good resource for obtaining up-to-date salary information for college and university faculty across the country.

Career Outlook

Overall, employment of college teachers is expected to grow as much as 27 percent through 2014. A significant portion of these new jobs will be part-time positions, however. Still, job opportunities are generally expected to be very good—although they will vary somewhat from field to field—as numerous openings for all types of postsecondary teachers result from the retirement of current teachers and a continued increase in student enrollment.

Ph.D. recipients seeking jobs as postsecondary teachers should experience favorable job prospects over the next decade. While competition will remain tight for tenure-track positions at four-year colleges and universities, a considerable number of part-time or renewable term appointments at these institutions should become available, in addition to various positions at community colleges. Opportunities for master's degree holders are also expected to be favorable since community colleges and other institutions that employ them, such as professional career education programs, are expected to experience considerable growth.

Opportunities for graduate teaching assistants should also be very good due to the likelihood of much higher undergraduate enrollments, coupled with more modest graduate enrollment increases. Constituting almost nine percent of all postsecondary teachers, graduate teaching assistants play an integral role in the postsecondary education system, and are expected to continue to do so in the future.

Advancement Possibilities

Standard advancement opportunities exist at most colleges and universities. Faculty members are promoted from assistant professor to associate professor to full professor. In addition, administrative positions are available for those who wish to follow this path. In some cases, these positions are viewed as promotions, while in others they are seen as temporary appointments. After holding one of these positions temporarily, the faculty member then returns to his or her academic department to resume his or her academic career of teaching, research, and/or extension.

The following list demonstrates the range of opportunities that exist for botany and horticulture faculty:

- Postdoctoral fellow
- Adjunct professor
- Research associate
- Lecturer
- Assistant professor, tenure track or nontenure track
- Associate professor, tenure track or nontenure track
- Full professor, usually only tenured but very occasionally nontenured
- Program director
- Department chairperson
- Assistant dean
- Associate dean
- Dean
- Vice president of research, graduate studies, or extension

Professional Associations

Almost every professional society in botany and horticulture represents the interest of its faculty members as well as its members in business, industry, and government. See Appendix B for a complete listing of professional societies and associations.

C H A P T E R
6

GOVERNMENT OPPORTUNITIES FOR BOTANISTS AND HORTICULTURISTS

Many botanists and horticulturists are employed by government agencies at all levels: federal, state, and local. Jobs with such offices can present excellent opportunities for challenging and fulfilling careers. In addition, they offer a wide variety of career paths for botanists and horticulturists.

FEDERAL GOVERNMENT JOBS

According to the U.S. Office of Personnel Management (OPM), the federal government hires between 200,000 and 300,000 people annually. A good number of these are botanists and horticulturists, and are employed in such agencies as the National Park Service, the U.S. Department of Agriculture, and the Environmental Protection Agency. The work they do and the background and experience required vary from discipline to discipline and agency to agency.

An interesting aspect of employment with government agencies is that they often involve dealings with private sector companies, colleges and universities, and foreign governments. This range of interaction can provide numerous opportunities and the possibility of a more diverse and varied career path.

Government careers can also be very satisfying. Compensation for botanists and horticulturists employed by government agencies tends to be competitive with salaries found in the private sector. In addition, government employment provides good fringe benefits, including health insurance and paid time off.

The work performed in federal agencies is often on the cutting edge, particularly in terms of research and development. For example, consider the National Science Foundation (NSF). The NSF is an independent agency of the federal government that is committed to promoting and supporting research and education programs in science and engineering. To accomplish its goals, the NSF establishes cooperative agreements with universities, consortia, and nonprofit research organizations in order to fund major research projects that advance knowledge in a particular field. For instance, research may be funded in the area of plant genetics or hydroponics in order to develop stronger varieties of food crops to feed people in third-world countries.

Other federal agencies, such as the National Park Service and the Department of Agriculture, hire far more botanists and horticulturists than the NSF.

The National Park Service is part of the U.S. Department of the Interior, which is the largest conservation agency in the country. Because the latter is responsible for the preservation of public lands and natural resources, the National Park Service is a vital component of the agency and offers opportunities to plan, design, implement, and preserve such areas as historical and botanical gardens, national parks, and the grounds of historic monuments, work which is carried out in large and small communities throughout the United States.

The Department of Agriculture, on the other hand, is responsible for helping the country's farmers and ranchers by implementing agricultural policies to maintain and improve farm income; develop and expand foreign markets for American agricultural products; prevent poverty, hunger, and malnutrition; enhance productivity and help the environment through conservation; promote rural development; conduct research; and maintain the quality and safety of the food supply.

This work is done by the divisions of Small Community and Rural Development; Marketing and Inspection Services; Food and Consumer Services; International Affairs and Commodity Programs; Science and Education; Natural Resources and Environment; and Economics. The diversity of tasks performed by the Department of Agriculture provides a wide variety of career opportunities for botanists and horticulturists.

Work by the agency is not just carried out in Washington, D.C. In fact, 90 percent of it is performed by its numerous county and district offices throughout the country, as well as its offices overseas. Because of this, the

Department of Agriculture hires a wide variety of professionals to accomplish its goals, many of whom are horticulturists and botanists.

Federal employees are paid according to a general schedule (GS) scale. The GS scale runs from grades 1 through 15; each grade having ten steps in which a salary incrementally increases. Most botanists and horticulturists would likely start at the GS-5 level or higher. The following are the 2006 GS salaries:

GS Level	Step 1	Step 5	Step 10
GS-5	$25,195	$28,555	$32,755
GS-6	28,085	31,829	36,509
GS-7	31,209	35,369	40,569
GS-8	34,563	39,171	44,931
GS-9	38,175	43,267	49,632
GS-10	42,040	47,644	54,649
GS-11	46,189	52,349	60,049
GS-12	55,360	62,740	71,965
GS-13	65,832	74,608	85,578
GS-14	77,793	88,165	101,130
GS-15	91,507	103,707	118,957

The Future Outlook for Federal Employment

Despite budget deficits and a general slowdown in government hiring, the Office of Management and Budget predicts that the demand for botanists and horticulturists will be higher than average because of new areas of biotechnology and the increased concern over environmental issues.

While most horticulturists' and botanists' jobs are in the executive branch of the federal government (for instance, the Department of Agriculture and the Department of the Interior), opportunities also exist in the legislative branch (for example, the U.S. Botanic Garden) and in independent agencies and government corporations such as the Environmental Protection Agency or the National Science Foundation.

STATE AND LOCAL GOVERNMENT JOBS

Many opportunities are available for botanists and horticulturists to work at state and local levels, given that state departments of agriculture have traditionally been principal employers.

Usually competition for state and local government jobs is not as stiff as it is at the federal level or in the private sector. However, due to regional and local differences, the career progression for horticulturists and botanists may not be as uniform as it is on the federal level. Therefore, it is important to educate yourself about the career paths in the locality where you hope to work.

Two resources can help you determine the career path of such vocations at the state and local levels. Those resources are

Council of State Governments
2760 Research Park Drive
P.O. Box 11910
Lexington, KY 40578-1910
csg.org

This resource is organized by state and updated biennially. It also provides the names and addresses of officials in each type of state administrative department with whom you can correspond to learn about career paths for employees in your particular field of botany or horticulture.

Municipal Year Book
International City/County Management Association
777 North Capitol Street NE, Suite 500
Washington, DC 20002
icma.org

This publication lists the addresses of city and county administration offices and departments and is updated annually. Again, it is an excellent resource for identifying professionals who can offer insight into the career paths available in local government agencies.

MATCHING YOUR INTERESTS TO OPPORTUNITIES IN GOVERNMENT

If you have an interest in resource management, the U.S. Department of the Interior provides numerous opportunities in land and water reclamation. The Bureau of Reclamation is the government's principal water development and management agency. Visit www.usbr.gov for more information.

The U.S. Geological Survey investigates and assesses land and water, conducts research on global change, and investigates natural hazards. Information is available at www.usgs.gov. Likewise, the Bureau of Land Management is responsible for nearly 300 million acres of public lands. The resources managed include rangeland vegetation, endangered plant species, and various wilderness areas. Visit www.blm.gov for details.

If you're interested in promoting technical and scientific knowledge about plants, the Smithsonian Institution provides the opportunity for botanists and horticulturists to present exhibits, conduct research, publish studies, and participate in international programs of scholarly exchange. Information about the various sites of the Smithsonian is available at www.si.edu.

For those interested in international work as well as horticulture or botany, the Peace Corps offers an unprecedented opportunity for you to share your knowledge with the world. The Peace Corps recruits more than 3,000 people annually to promote world peace, friendship, and understanding in over 130 countries. For periods of two to three years or longer, volunteers work in a variety of entry-level positions helping governments and their people learn to grow, propagate, and harvest sustainable crops to improve the quality of life in those countries. Horticulture and botany professionals are in high demand by the Peace Corps. Visit www.peacecorps.gov for detailed information about volunteer opportunities.

If you have an interest in our country's space program, the National Aeronautics and Space Administration (NASA) offers opportunities to address such cutting-edge issues as feeding astronauts in space. NASA has been working to identify plant materials that can be used to sustain life in the final frontier. Through five university centers around the country, NASA has funded faculty to research and develop strategies for growing, processing, and preparing food in outer space.

The NASA Center for Food Production, Processing, and Waste Management in Controlled Ecological Life Support Systems (CELSS) is one such project. The CELSS project uses the expertise of plant physiologists, plant nutritionists, plant pathologists, horticulturists, microbiologists, plant breeders, plant geneticists, plant modelers, biochemists, food chemists, food scientists, soil scientists, and water chemists, as well as numerous engineers and computer scientists. The overall goal of the CELSS project is to provide tested information and technologies applicable to bioregenerative food production systems for life support on long-term manned space missions. Specifically, the Center is developing information, computer-simulated models, methodologies, and technology for sweet potato and peanut biomass production and processing. This work also includes waste management and the recycling of those crops selected by NASA for CELSS.

The Center is organized into interdisciplinary teams of life scientists and engineers that work together on specific objectives and long-term goals. Integral to its goal is the development of both basic and applied research information. In addition, the Center is dedicated to the training of young scientists and engineers, especially underrepresented minorities that will increase the professional pool in these disciplines and contribute to the advancement of space sciences and exploration.

The Center serves as a focal point for integrating research on subsurface crops within the program. Research there falls under one of four subsystems:

- Biomass production, which includes environmental control and the computer modeling of plant growth in CELSS conditions
- Sweet potato and peanut germplasm development for CELSS conditions
- Nutrition and processing, which includes nutrient analysis and the preparation of a variety of foods from these crops
- Recycling of waste from biomass production and processing

The Center is composed of six functional working groups that interface on a continuous basis. The targeted goals of each working group within the Center are summarized in the following:

- *Growing Systems and Environmental Factors (GRO)*: To ascertain the best systems and environmental conditions for growing sweet potato and peanut hydroponically for space missions
- *Microgravity Applications and Controls Group (MAC)*: To adapt efficient crop producing hydroponic systems for use under microgravity conditions and to design and build automated environmental control systems for crop growing and waste management systems
- *Plant Modeling Group (PAM)*: To establish a database from existing information, recommend and design experiments for

retrieving needed information and, based on these data, produce models that effectively predict growth and yield of sweet potatoes and peanuts in CELSS

- *Nutrition and Food Processing Group (NAF)*: To analyze the nutrient composition of the edible parts (greens and roots/nuts) of plants grown under controlled environmental conditions and to process these parts into a variety of nutritious and palatable foods
- *Germplasm Development Group (GED)*: To examine the germplasm available for peanut and sweet potato; to select and breed the best lines for growing in CELSS and to further improve them through biotechnology
- *Waste Management and Recycling Group (WAM)*: To ascertain quantities of inedible plant biomass available from sweet potato and peanut in CELSS, to analyze the biomass for chemical composition, and to establish how all organic and inorganic waste resources from the hydroponic culture of these crops will be recovered and recycled in CELSS

Complete information about NASA programs is available at www.nasa.gov.

If you're interested in intelligence work, numerous government agencies, such as the National Security Agency (www.nsa.gov) or the Central Intelligence Agency (www.cia.gov), offer an opportunity for people with backgrounds in botany and horticulture. In the intelligence world, professionals in these areas work with state-of-the-art technology to identify indigenous plant materials around the world. This provides vital information about environmental conditions that may impact the quality of life of the world community.

SPECIAL TIPS

Participation in a college or university cooperative education (co-op) program is one of the best ways to secure a permanent position with a government agency. As a co-op student with a federal agency, you're considered to be an employee of the agency, and upon graduation are eligible to convert your co-op employment to permanent employment without competition, if full-time positions are available. The real benefit of this conversion is that your seniority is dated from your original employment date as a co-op student. This has a positive impact on your eligibility for certain federal benefits, including sick leave, vacation, and ultimately retirement. See your school's co-op office about applying to federal, state, and local agencies in which you are interested.

Likewise, special internship programs allow students to gain experience with a government agency prior to graduation. Internships are typically shorter in duration than co-op experiences; however, they provide good opportunities to learn about the nature of work and the career opportunities available in federal, state, and local government agencies.

Participants in both co-op and internship programs generally are selected on the basis of academic performance and demonstrated leadership skills. It is important to use the resources of your college or university in preparing to compete for these positions. A strong resume and good interviewing skills are essential in being able to secure an internship or co-op position.

C H A P T E R

7

INDUSTRY OPPORTUNITIES FOR BOTANISTS AND HORTICULTURISTS

Industry provides many botanists and horticulturists with a wide and diverse range of career paths. While the opportunities are always changing in terms of how botanists and horticulturists work in industry, there is no indication it will lose its position as an important place of employment for those in these career fields.

Private industry employs ten percent of botanists. The types of industries most likely to hire people in this field are drug companies, the oil industry, the chemical industry, lumber and paper companies, seed and nursery companies, fruit growers, food companies, fermentation industries (including breweries), biological supply houses, and biotechnology firms.

It is not necessary for botanists and horticulturists to begin their careers in industry in order to have a successful career path in the private sector. For example, some botanists and horticulturists who initially join federal or state regulatory agencies and gain several years of experience find it easy to move into industrial positions.

SOURCES OF INFORMATION ON BOTANY AND HORTICULTURE CAREERS

Every year, many people seek to begin careers in botany and horticulture. They often become frustrated in their job search, however, because many typical sources of job leads don't list organizations seeking workers. This

125

doesn't mean such sources don't exist. On the contrary, a number of good resources can be used to search for summer jobs, internships, cooperative education opportunities, and permanent positions. These resources include, but are not limited to, the following:

- The Research Centers Directory
- The Directory of American Agriculture
- Cooperative State Research, Education, and Extension Service (CSREES) (Visit www.csrees.usda.gov for contact information on all local extension offices across the country.)
- The Encyclopedia of Associations (This lists almost every professional association in existence today.)
- The Office of Personnel Management web site (Visit www.usajobs.com for an online listing of all available federal jobs.)
- College and university alumni associations where biology, botany, and horticulture degrees are granted

SOME POTENTIAL EMPLOYERS OF BOTANY AND HORTICULTURE PROFESSIONALS

The following is a list of potential employers of botanists and horticulturists in the agricultural, paper and lumber, and pharmaceutical industries.

Agricultural Industries

Archer Daniels Midland Co.
4666 Faries Parkway
P.O. Box 1470
Decatur, IL 62526
admworld.com

Birdsong Corp.
612 Madison Avenue
Suffolk, VA 23434-4028

Castle and Cooke, Inc.
10900 Wilshire Boulevard
Los Angeles, CA 90024
castlecooke.net

ConAgra Foods, Inc.
conagrafoods.com

Corn Products International, Inc.
5 Westbrook Corporate Center
Westchester, IL 60154
cornproducts.com

Curtice Burns Foods, Inc.
90 Linden Place
Rochester, NY 14625

Del Monte Foods Co.
delmonte.com

Dole Food Co., Inc.
dole.com

Duda and Sons, Inc.
duda.com

Flowers Industries, Inc.
P.O. Box 4397
Spartanburg, SC 29305
foodpros.com

General Mills, Inc.
P.O. Box 9452
Minneapolis, MN 55440
generalmills.com

Grupo Industrial Maseca Sa de
CV Gimsa
2520 Avenida La Clinica Colonia
Sertoma
Monterrey 64710, Nuevo Leon
64710
Mexico

Hanover Foods Corp.
P.O. Box 334
Hanover, PA 17331
hanoverfoods.com

Heinz Foods
P.O. Box 57
Pittsburgh, PA 15230
heinz.com

Holly Farms Corp.
1755-D Lynnfield Road,
Suite 149
Memphis, TN 38119

Imperial Sugar
P.O. Box 4225
Savannah, GA 31407-4225
imperialsugar.com

J. M. Smuckcr Co.
One Strawberry Lane
Orville, OH 44667-0280
smuckers.com

J. R. Simplot Foods, Inc.
999 Main Street, Suite 1300
Boise, ID 83702
simplot.com

Kaliroy Pacific
Nogales, AZ 34221
sunripeproduce.com

Kellogg Co.
One Kellogg Square
P.O. Box 3599
Battle Creek, MI 49016-3599
kelloggs.com

Kraft Foods
kraftfoods.com

Lancaster Colony Corp.
37 West Broad Street
Columbus, OH 43215
lancastercolony.com

Maui Land and Pineapple
    Co., Inc.
P.O. Box 187
Kahului, HI 96733-6687
mauiland.com

Pacific Collier Fresh
925 New Harvest Road
Immokalee, FL 34142
sunripeproduce.com

Pacific Tomato Growers, Inc.
503 Tenth Street West
Palmetto, FL 33221
sunripeproduce.com

Pacific Triple E
8690 West Linne Road
Tracy, CA 95304
sunripeproduce.com

Pilgrims Pride Corp.
P.O. Box 93
Pittsburg, TX 75686
pilgrimspride.com

Purina
purina.com

Ralcorp Holdings, Inc.
P.O. Box 618
St. Louis, MO 63118-0618
ralcorp.com

Riviana Foods, Inc.
riviana.com

Seneca Foods Corp.
3736 South Main Street
Marion, NY 14505
senecafoods.com

Stokely USA, Inc.
1055 Corporate Center Drive
Oconomowoc, WI 53066

Sunsweet Growers, Inc.
901 North Walton Avenue
Yuba City, CA 95993
sunsweet.com

Tomkins PLC
East Putney House
84 Upper Richmond Road
London SW15 2ST
United Kingdom
tomkins.co.uk

United Brands
7 West 41st Avenue, #512
San Mateo, CA 94403
unitedbrands.us

United Natural Foods, Inc.
P.O. Box 999
260 Lake Road
Dayville CT 06241
unfi.com

Universal Foods Corp.
433 East Michigan Street
Milwaukee, WI 53202-5106

Paper and Lumber Industries

Albany International Corp.
P.O. Box 1907
Albany, NY 12201-1907
albint.com

American Cyanamid Co.
4201 Quakerbridge Road
Princeton Junction, NJ 08550

American Greetings Corp.
One American Road
Cleveland, OH 44144-2398
http://corporate.americangreetings.
  com

Anheuser Busch Cos., Inc.
One Busch Place
St. Louis, MO 63118
anheuser-busch.com

Asia Pacific Resources Intl.
  Holdings Ltd.
Singapore
april.com.sg

Asia Pulp and Paper Co. Ltd.
Singapore
asiapulppaper.com

Avenor, Inc.
1250 Rene Levesque Boulevard
  West
Montreal, Quebec
Canada H3B 4Y3

Boise Cascade Corp.
1111 West Jefferson Street
P.O. Box 50
Boise, ID 83728
bc.com

Bowater, Inc.
bowater.com

Chemed Corp.
2600 Chemed Center
255 East Fifth Street
Cincinnati, OH 45202-4726
chemed.com

Corporacion Durango
Potasio 150
Cuidad Industrial
Durango, Dgo.
Mexico
corpdgo.com

Cytec Industries, Inc.
Five Garret Mountain Plaza
West Paterson, NJ 07424
cytec.com

Du Pont de Nemours and Co.
1007 Market Street
Wilmington, DE 19898
dupont.com

Engelhard Corp.
101 Wood Avenue
Iselin, NJ 08830
engelhard.com

Falconbridge Ltd.
207 Queen's Quay West, Suite 800
Toronto, ON, Canada M5J 1A7
noranda.com

Federal Paper Board Co., Inc.
75 Chestnut Ridge Road
Montvale, NJ 07645

Fletcher Building Ltd.
New Zealand
fletcherbuilding.co.nz

Fort Howard Steel
1240 Contract Drive
Green Bay, WI 54304

Georgia-Pacific Corp.
gp.com

Gibson Greetings, Inc.
2100 Section Road
Cincinnati, OH 45237

Hercules, Inc.
1313 North Market Street
Wilmington, DE 19894-0001
herc.com

IKON Office Solutions
70 Valley Stream Parkway
Malvern, PA 19355-0989
ikon.com

International Paper Co.
400 Atlantic Street
Stamford, CT 06921
internationalpaper.com

J. B. Hunt Transport Services, Inc.
P.O. Box 130
615 J. B. Hunt Corporate Drive
Lowell, AR 72745
jbhunt.com

Johns Manville Corp.
P.O. Box 5108
Denver, Colorado 80217-5108
jm.com

Kimberly-Clark Corp.
Dept. INT
P.O. Box 2020
Neenah, WI 54957-2020
kimberly-clark.com

Mead Westvaco
One High Ridge Park
Stamford, CT 06905
meadwestvaco.com

Nash Finch Co.
7600 France Avenue South
Edina, MN 55435
nashfinch.com

New York Times Co.
229 West 43rd Street
New York, NY 10036
nytco.com

Rykoff Sexton, Inc.
613 Baltimore Drive
Wilkes-Barre, PA 18702

Sequa Corp.
sequa.com

Smurfit-Stone Container Corp.
150 North Michigan Avenue
Chicago, IL 60601
smurfit-stone.com

Sonoco Products Co.
sunoco.com

St. Joe Paper Co.
245 Riverside Avenue,
    Suite 500
Jacksonville, FL 32202
joe.com

Sweetheart Holdings, Inc.
10100 Reisterstown Road
Owings Mills, MD 21117

Swift Transportation Co., Inc.
2200 South 75th Avenue
Phoenix, AZ 85043
swifttrans.com

Sysco Corp.
1390 Enclave Parkway
Houston, TX 77077-2099
sysco.com

Valhi, Inc.
5430 LBJ Freeway
Dallas, TX 75240-2697

Werner Enterprises, Inc.
P.O. Box 45308
Omaha, NE 68145-0308
werner.com

Weyerhaeuser Co.
P.O. Box 9777
Federal Way, WA 98063-9777
weyerhaeuser.com

Xerox Corp.
xerox.com

Pharmaceutical Industries

Abbott Laboratories
Abbott Park, IL 60064
abbottdiagnostics.com

American Cyanamid Co.
4201 Quakerbridge Road
Princeton Junction,
    NJ 08550

Baxter Healthcare
One Baxter Parkway
Deerfield, IL 60015-4625
baxter.com

Bristol Myers Squibb Co.
345 Park Avenue
New York, NY 10154-0037
bms.com

Eli Lilly and Co.
Lilly Corporate Center
Indianapolis, IN 46285
lilly.com

GlaxoSmithKline
gsk.com

Ivax Pharmaceuticals
4400 Biscayne Blvd.
Miami, Florida 33137
ivaxpharmaceuticals.com

Marion Merrell Dow, Inc.
9300 Ward Parkway
Kansas City, MO 64114

Merck and Co., Inc.
One Merck Drive
P.O. Box 100
Whitehouse Station, NJ
    08889-0100
merck.com

Novo Nordisk AS
Krogshoejvej 41
DK-2880 Bagsvaerd
Denmark
novo.dk

Pfizer, Inc.
235 East 42nd Street
New York, NY 10017
pfizer.com

Rhone Poulenc Rorer, Inc.
P.O. Box 1200
500 Arcola Road
Collegeville, PA 19426-9957

Sigma Aldrich Corp.
3050 Spruce Street
St. Louis, MO 63103
sigmaaldrich.com

Teva Pharmaceuticals
5 Basel Street
Petach Tivka 49131
Israel
tevapharm.com

Valeant Pharmaceuticals
    International
3300 Hyland Avenue
Costa Mesa, CA 92626
valeant.com

Wyeth
5 Giralda Farms
Madison, NJ 07940
wyeth.com

C H A P T E R

8

PLANNING FOR GRADUATE SCHOOL

If your career in botany or horticulture requires an advanced degree, your selection of an appropriate graduate school is very important. First and foremost, you will want to pursue your graduate study at an institution that best prepares you for the type of career you're seeking. However, you will also want to study at an institution that best meets your personal needs in terms of such things as curriculum requirements, geographic location, and personal cost versus financial aid availability.

Although top-ranked institutions usually provide the best platform for almost any career, admission to such programs is not enough. You must perform well and earn the support of your major's professors in order to gain the most from your graduate education. Over the long term, the reputation of the university you select and the recommendations of your graduate faculty advisor may help open doors and assure you success in your career. Therefore, when possible and appropriate, graduate schools consistently ranked among the top 25 programs in your field are excellent institutions to consider. The reputations of these schools, the quality of their faculty, and your personal credentials will make you more competitive in a tight job market.

IDENTIFYING THE RIGHT GRADUATE SCHOOL FOR YOU

Numerous directories are available to help you identify graduate schools in your area of interest. These provide information on such things as the faculty/student ratio (it should be low), the number of master's and doctoral degrees awarded annually, the percentage of women and minorities admitted, whom to contact for application materials, and the list of standardized tests required for admission.

Publications can also be found that rank graduate programs. *The Princeton Review* rates domestic and international graduate and professional programs, and *U.S. News and World Report* annually ranks U.S. graduate schools and specific programs. These directories are useful tools when making a decision about which graduate programs you wish to apply to.

In using these rankings, it's advisable to examine the factors employed to rank the schools, as some factors may be more important to you than others. For example, if you are planning to obtain a Ph.D., it's important to know how the department or school ranks among other institutions, and how it ranks in terms of faculty publications. However, if you plan to enter private industry after graduation, it's more important to know how industry rates the department or school to which you plan to apply.

It's also advisable to determine if the program of study is ranked separately from the school or university in general. If it is, you should determine what the departmental ranking is in relation to the school/college or the university. A good rule-of-thumb when trying to make this determination is to assess whether or not the ranking of your intended department is as high or higher than the overall ranking of the school or university. For example, you would want to determine if the Horticulture Department is ranked as highly as the School/College of Agriculture and/or the University. If it is equally ranked or ranked higher, the quality of the department is probably indisputable. If it is ranked lower, you will want to ask probing questions of the faculty and graduate students at that institution before making the decision to apply and/or attend.

Another excellent resource for information about graduate programs in your area of interest is members of your undergraduate faculty. Some students who plan to go on to graduate school fail to seek the advice and guidance of undergraduate faculty when making decisions about graduate school, which can be a significant oversight. University and college faculty are generally aware of their counterparts at other universities, and are usually knowledgeable about the quality and quantity of the research of faculty at other institutions. They also tend to be aware of other faculty members' reputations in the field.

The input of faculty can be particularly helpful when you are in the process of completing the graduate school application form. Not only can these faculty provide you with letters of recommendation that address your academic performance and potential, they can also assist you in developing a well-targeted "Statement of Purpose," which is generally required on all graduate school applications.

Unlike admissions to undergraduate school, graduate applications are reviewed and evaluated by faculty members in the department in which you plan to study. Thus, before you apply to graduate school, it's advisable to talk to the faculty in your undergraduate department and learn all you can about the faculty in the department to which you plan to apply. For instance, some good questions to ask might be

- Where did they earn their degrees?
- What is their area of expertise?
- What books and articles have they published?
- Are their professional interests similar to yours?

The following suggestions will help you assess whether the graduate faculty's research interests are similar to yours.

- Consult resources such as the *Peterson's Guide*, available at www.petersons.com. The site allows users to search for schools and programs based on several criteria, providing detailed information about faculty members and their research.
- Conduct literature searches to locate research papers faculty members have written and to also become more familiar with the field.
- Call the department and inquire about their active areas of research. Follow up with telephone calls or e-mail messages to the faculty members who are active researchers and ask about their work. It is important for you to determine if their area of interest matches yours. You also want to assess whether you think you could work and study with this person. Any contact like this should end with a request for information about other universities and colleagues working in the same research area.
- Visit schools and tour the department in which you are interested. This is the best way to evaluate any school or program. How do you feel when you're on the campus? In the department? Talking to the faculty? Talking to current graduate students?

Key issues in deciding where to apply for graduate study include

- A good match between your research interests and the department's strengths
- The reputation of the department and/or the specialization in which you are interested
- The overall quality of the institution

APPLYING TO GRADUATE SCHOOL

Most graduate programs require the Graduate Record Examination (GRE). It is advisable to take the GRE during your senior year of college. Even if you do not plan to go to graduate school immediately after completing your undergraduate degree there are three principal reasons for taking the GRE during senior year.

First, as an undergraduate, you are used to taking tests and exams. Second, while you are in school, your general knowledge, as well as knowledge of your specific discipline, is current and up-to-date. Third, the scores are good for five years, and having taken the exam, all of your options for the future remain open. It can be intimidating to take this exam when you have not been in school for several years!

The next step in the process is the completion of the graduate application. The basic steps include

- Completing the application form
- Submitting official transcripts
- Providing three letters of reference
- Writing a statement of purpose
- Requesting financial assistance

FINANCING YOUR GRADUATE EDUCATION

Paying for graduate school should not be a barrier to an academic career. If you have a strong record of academic achievement, you should be competitive for numerous fellowships and graduate assistantships. These not only pay your tuition and fees but also provide a monthly stipend for living expenses. Some of these sources include

- *Teaching assistantships*, which provide stipends, and sometimes tuition waivers, to full-time graduate students for assisting a faculty member in teaching undergraduate classes
- *Research assistantships*, which provide stipends, and sometimes tuition waivers, to full-time graduate students for assistance on a faculty research project
- *Fellowships*, which provide money to full-time graduate students to cover the costs of study and living expenses and are not based on an obligation to assist in teaching or conducting research. Some fellowships are funded by the school, but others are available from outside sources, such as the National Science Foundation. These fellowships provide an excellent credential when applying for post-doctoral positions, and ultimately faculty positions, because they are so competitive. If these sources of funding are not available to you, consider applying for student loans. The government maintains a web site where you can get complete information about student loans. Visit www.nslds.ed. gov for details.

In seeking funding for your graduate education, it is highly recommended that you do your homework. Talk to faculty members and graduate schools, and learn as much as you can about the process of competing for fellowships and assistantships. If you prejudge your own competitiveness, you may be making a costly mistake and incurring unnecessary debt!

It is important to have a good undergraduate record and evidence of commitment to the field. This commitment can be in the form of paid, related work experience, or volunteer experience. The faculty committee that reviews applications and makes recommendations for awarding financial assistance needs to see that you can make a substantive contribution to their teaching and/or research.

A list of universities and colleges in the United States and Canada offering degrees in agriculture are given in Appendix A.

CHAPTER 9

MENTORING TIPS FROM HORTICULTURE AND BOTANY PROFESSIONALS

Unlike those who work in such fields as accounting, law, engineering, or teaching, it can be difficult to readily identify people who work in the various botany and horticulture occupations. So if you have questions about botany and horticulture, where do you go to get advice?

In this chapter, we hope to resolve some of that problem by providing you with an opportunity to hear directly from professionals who work in the many areas of botany and horticulture. The people interviewed were asked to describe specific factors that would be helpful to the readers of this book. The points they were asked to discuss include the following:

- Why they chose their career field
- What tasks they perform on a regular basis
- What tips they would give to others who are considering their occupational field

The following personal accounts of three professionals should help give you some insight into what to expect in a career in horticulture or botany.

LISA: PLANT SCIENCE TECHNICIAN

With an interest in various aspects of science, Lisa chose plant science because it does not require the use, or killing, of animals. She prefers working in research as opposed to industry because it offers an opportunity for continual learning. In terms of career advancement, research generally involves more teamwork and less competition than industry. In addition, those who work in applied research are able to see immediate results of their work.

The goals of Lisa's work are two-fold. She conducts research on various crops to help solve growers' problems and try to find an explanation for how systems work. The results of her research may be immediately useful to a grower or may provide a basis for future research.

Lisa has worked with genetic crosses, variety trials, and the detailed examination of a biocontrol fungus. Her daily duties range from cleaning a greenhouse to harvesting a field to chemiluminescence. One of the things Lisa enjoys most about her position is the variety of tasks she gets to perform. Her job involves greenhouse work, laboratory testing, fieldwork, and computer analysis of data.

As new experiments arise, Lisa has the opportunity to use different types of equipment such as microscopes, centrifuges, autoclaves, electrophoretic equipment, spectrophotometers, planters, rototillers, growth chambers, and even unique equipment designed specifically for particular experiments.

For those considering a career in plant science, Lisa suggests that once you've taken the basic courses required for a major in biology or botany, you should next take those elective classes that interest you, rather than concentrating on the ones that will make you more marketable. As she says, "While it is good to have classes that will sell you to your first employer, you never know what field will be most employable in a few years. It is best to try to make the extra effort to find a job that will be fun for you."

Lisa also recommends finding summer jobs in greenhouses. An independent study project can offer significant research experience in this area. An alternative, less formal approach is to ask a favorite professor if you can volunteer on a project. Even mundane tasks can provide valuable experience.

NORMAN: PLANT GENETICIST AND ASSOCIATE PROFESSOR

As he received a bachelor of science degree in chemistry, Norman realized he was more interested in biological systems than chemistry. He went to graduate school and studied stress in rabbit populations, earning a master's degree in biology. He hadn't planned to be a horticulturist at the time, but following his interests and the job market led to his career as a plant geneticist working with horticultural crops.

Norman's first job was with an engineering firm, where he was hired to write environmental impact statements. He later worked for a civil engineering firm as supervisor of the company's water quality lab, where he found himself back doing chemistry. During this period, Norman also wrote a book on the flora of the Sierra Nevada.

He later decided to pursue a Ph.D. in botany, and realized that the genetics program was what interested him most. With this in mind, he graduated in plant genetics with an emphasis in evolution, and secured a postdoctoral position in plant biochemistry/evolution. Following his work as a post doc, Norman accepted his current faculty position. The job was originally defined as a plant biochemist working with seeds, but has since been redefined according to Norman's interests and developments in the field of plant science. He is now a plant geneticist working with breeders.

When asked what advice he could offer potential plant geneticists, Norman recommended that students follow their interests while keeping an eye on possible alternative employment opportunities.

HELENE: PLANT PATHOLOGIST, ASSOCIATE PROFESSOR, AND EXTENSION FACULTY MEMBER

Helene is employed as an associate professor of plant pathology, specializing in vegetable pathology. Her work involves 50-percent research and 50-percent extension and outreach on vegetable diseases, involving her with the commercial agricultural community.

Helene majored in the biology of natural resources at the University of California—Berkeley, planning to pursuing a career in forestry or with the National Park Service. She went on to earn a master's degree in soil science with a concentration in plant nutrition, then took civil service exams in the hope of securing a forestry position. However, the grading system for the exams allows additional points for military veterans, and without these points Helene's scores were not high enough, even though they were in the 90s.

With forestry no longer an option, Helene began to think about careers in agriculture. She says, "Plant pathology had been my absolute favorite course as an undergraduate student. So I went with my heart." Following her instincts, she entered a Ph.D. program in plant pathology, which gave her the opportunity to work in the lab of a famous vegetable pathologist, where she gained a great deal of valuable experience. Immediately after graduation, Helene was hired for a faculty position at Cornell University.

Currently, Helene works with farmers to help them grow healthy vegetables that can be canned, frozen, or sold as fresh produce. She identifies bacteria and fungi that cause disease, and looks for ways to stop such diseases and prevent others from forming. Her primary work is with fungi, studying their interactions with plants. To accomplish this, Helene uses special equipment such as scanning electron microscopes and pure air stations created by laminar flow hoods.

Helene's advice for anyone considering a career in plant pathology is to read as much as possible across all disciplines. If you're interested in a particular subject, she suggests you apply to work in a laboratory that focuses on that area. In Helene's case, she began by washing dishes in a soil scientist's lab, where she learned a lot about the way a scientific study is conducted. Her next work study job was growing strawberries in a greenhouse for a researcher studying plant nutrition. She explains the benefit of these experiences by saying, "I realized I was 'doing chemistry' without taking a boring course—and that made me work harder in the chemistry courses I did take."

As a university professor, Helene now hires high-school and college students on a part-time basis so they can learn first-hand what goes on in a lab. During the summer months, they work full-time. Perhaps as a result, many of her former employees have since gone on to high-paying jobs in the sciences.

APPENDIX

A

UNIVERSITIES AND COLLEGES IN THE UNITED STATES AND CANADA OFFERING DEGREES IN AGRICULTURE

Alabama

Alabama A&M University
Normal, AL 35762
aamu.edu

Auburn University
Auburn, AL 36849-5122
grad.auburn.edu

Tuskegee University
Tuskegee, AL 36088
tuskegee.edu

Alaska

University of Alaska, Fairbanks
Fairbanks, AK 99775
uaf.edu

Arizona

Arizona State University
Tempe, AZ 85281
asu.edu

University of Arizona
Tucson, AZ 85721
arizona.edu

Arkansas

University of Arkansas
Fayetteville, AR 72701
uark.edu

University of Arkansas
Monticello, AR 71656
uamont.edu

University of Arkansas
Pine Bluff, AR 71601
uapb.edu

California
California Polytechnic State
    University
San Luis Obispo, CA 93407
calpoly.edu

California State Polytechnic
    University
Pomona, CA 91768
csupomona.edu

California State University, Chico
Chico, CA 95929
csuchico.edu

California State University,
    Fresno
Fresno, CA 93740-8027
csufresno.edu

Humboldt State University
Arcata, CA 95521-8929
humboldt.edu

University of California
Davis, CA 95616
ucdavis.edu

University of California,
    Riverside
Riverside, CA 92521
ucriverside.edu

Colorado
Colorado State University
Fort Collins, CO 80523
http://welcome.colostate.edu

Fort Lewis College
Durango, CO 81301-3999
fortlewis.edu

Connecticut
University of Connecticut
Storrs, CT 06269
uconn.edu

Delaware
Delaware State University
Dover, DE 19901
desu.edu

University of Delaware
Newark, DE 19716
udel.edu

Florida
Florida Southern College
Lakeland, FL 33801-5698
flsouthern.edu

University of Florida
Gainesville, FL 32611
ufl.edu

Georgia
Abraham Baldwin Agricultural
    College
Tifton, GA 31793
http://stallion.abac.peachnet.edu

Berry College
Mount Berry, GA 30149
berry.edu

Fort Valley State University
Fort Valley, GA 31030
fsvu.edu

University of Georgia
Athens, GA 30602
uga.edu

Hawaii
University of Hawaii
Honolulu, HI 96817
hawaii.edu/campuses/honolulu.
  html

Idaho
Brigham Young University
Rexburg, ID 83460
byui.edu

College of Southern Idaho
Twin Falls, ID 83303
csi.edu

University of Idaho
Moscow, ID 83844
uihome.uidaho.edu

Illinois
Illinois State University
Normal, IL 61761
ilstu.edu

Southern Illinois University
Carbondale, IL 62901
siu.edu

University of Illinois
Urbana, IL 61820
uiuc.edu

Western Illinois University
Macomb, IL 61455
wiu.edu

Indiana
Purdue University
West Lafayette, IN 47907
purdue.edu

Iowa
Iowa State University
Ames, IA 50011
iastate.edu

Kansas
Fort Hays State University
Hays, KS 67601
fhsu.edu

Kansas State University
Manhattan, KS 66506
k-state.edu

McPherson College
McPherson, KS 67460
mcpherson.edu

Kentucky
Morehead State University
Morehead, KY 40351
moreheadstate.edu

Murray State University
Murray, KY 42071
murraystate.edu

University of Kentucky
Lexington, KY 40506
uky.edu

Western Kentucky University
Bowling Green, KY 42101
wku.edu

Louisiana
Louisiana State University
Baton Rouge, LA 70803
lsu.edu

Louisiana Tech University
Ruston, LA 71272
latech.edu

McNeese State University
Lake Charles, LA 70609
mcneese.edu

Nicholls State University
Thibodaux, LA 70310
nicholls.edu

Northwestern State University
Natchitoches, LA 71497
nsula.edu

Southern University and A&M
  College
Baton Rouge, LA 70813
subr.edu

University of Louisiana
Lafayette, LA 70504
louisiana.edu

University of Louisiana
Monroe, LA 71209
ulm.edu

Maine
University of Maine
Orono, ME 04469
umaine.edu

Maryland
University of Maryland
College Park, MD 20742
umd.edu

University of Maryland Eastern
  Shore
Princess Anne, MD 21853
umes.edu

Massachusetts
University of Massachusetts
Amherst, MA 01003
umass.edu

Michigan
Michigan State University
East Lansing, MI 48824
msu.edu

Michigan Technological
    University
Houghton, MI 49931
mtu.edu

Northern Michigan University
Marquette, MI 49855
nmu.edu

Minnesota
University of Minnesota
    Crookston
Crookston, MN 56716
crk.umn.edu

University of Minnesota Twin
    Cities
St. Paul, MN 55108
www1.umn.edu/twincities

Mississippi
Alcorn State University
Alcorn State, MS 39096
alcorn.edu

Mississippi State University
Mississippi State, MS 39762
msstate.edu

Missouri
Central Missouri State University
Warrensburg, MO 64093
cmsu.edu

Lincoln University
Jefferson City, MO 65101
lincolnu.edu

Missouri Western State
    University
St. Joseph, MO 64507
missouriwestern.edu

Northwest Missouri State
    University
Maryville, MO 64468
nwmissouri.edu

Southwest Missouri State
    University
Springfield, MO 65897
missouristate.edu

University of Missouri
Columbia, MO 65211
missouri.edu

Montana
Montana State University
Bozeman, MT 59717
montana.edu

Montana State University—
    Northern
Havre, MT 59501
msun.edu

Nebraska
University of Nebraska
Lincoln, NE 68588
unl.edu

Nevada
University of Nevada
Reno, NV 89557
unr.edu

New Hampshire

University of New Hampshire
Durham, NH 03824
unh.edu

New Jersey

Rutgers University
New Brunswick, NJ 08903
rutgers.edu

New Mexico

New Mexico State University
Las Cruces, NM 88003
nmsu.edu

New York

Cornell University
Ithaca, NY 14853
cornell.edu

North Carolina

North Carolina A&T State
   University
Greensboro, NC 27411
ncat.edu

North Carolina State University
Raleigh, NC 27695
ncsu.edu

North Dakota

North Dakota State University
Fargo, ND 58105
ndsu.nodak.edu

Ohio

Ohio State University
Columbus, OH 43210
osu.edu

Wilmington College
Wilmington, OH 45177
wilmington.edu

Oklahoma

Cameron University
Lawton, OK 73505
cameron.edu

Langston University
Langston, OK 73050
lunet.edu

Oklahoma Panhandle State
   University
Goodwell, OK 73939
opsu.edu

Oklahoma State University
Stillwater, OK 74078
osu.okstate.edu

Oregon

Oregon State University
Corvallis, OR 97331
oregonstate.edu

Pennsylvania

Delaware Valley College of
   Science and Agriculture
Doylestown, PA 18901
devalcol.edu

Pennsylvania State University
University Park, PA 16802
psu.edu

Temple University
Ambler, PA 19122
temple.edu

Puerto Rico
University of Puerto Rico
Mayaguez, PR 00681
uprm.edu

University of Puerto Rico
Rio Piedras, PR 00928
upr.edu

Rhode Island
University of Rhode Island
Kingston, RI 02881
uri.edu

South Carolina
Clemson University
Clemson, SC 29634
clemson.edu

South Dakota
South Dakota State University
Brookings, SD 57007
www3.sdstate.edu

Tennessee
Austin Peay State University
Clarksville, TN 37044
apsu.edu

Middle Tennessee State
University
Murfreesboro, TN 37132
mtsu.edu

Tennessee Technological
University
Cookeville, TN 38505
tntech.edu

University of Tennessee
Knoxville, TN 37996
utk.edu

University of Tennessee
Martin, TN 38238
utm.edu

Texas
Abilene Christian University
Abilene, TX 79699
acu.edu

Prairie View A&M University
Prairie View, TX 77446-0519
pvamu.edu

Sam Houston State University
Huntsville, TX 77341
shsu.edu

Stephen F. Austin State
University
Nacogdoches, TX 75962
sfasu.edu

Texas A&I University
College Station, TX 77843
tamu.edu

Texas State University
San Marcos, TX 78666
txstate.edu

Texas Tech University
Lubbock, TX 79409
ttu.edu

Utah

Brigham Young University
Provo, UT 84602
byu.edu

Utah State University
Logan, UT 84322-0160
usu.edu

Vermont

University of Vermont
Burlington, VT 05405
uvm.edu

Virginia

Old Dominion University
Norfolk, VA 23529
odu.edu

Virginia State University
Petersburg, VA 23806
vsu.edu

Virginia Tech
Blacksburg, VA 24061
vt.edu

Washington

University of Washington
Seattle, WA 98195
washington.edu

Washington State University
Pullman, WA 99164-1067
wsu.edu

West Virginia

West Virginia University
Morgantown, WV 26506
wvu.edu

Wisconsin

University of Wisconsin
Green Bay, WI 54311-7001
uwgb.edu

Wyoming

University of Wyoming
Laramie, WY 82071
uwyo.edu

Canada

Dalhousie University
Halifax, NS B3H 4R2
dal.ca

McGill University
Montreal, Quebec H3A 2T5
mcgill.ca

Nova Scotia Agricultural College
Truro, NS B2N 5E3
nsac.ns.ca

Universite Laval
Sainte-Foy, Quebec G1K 7P4
ulaval.ca

University of Alberta
Edmonton, AB TG6 2E1
ualberta.ca

University of British Columbia
Vancouver, BC V6T 1Z4
ubc.ca

University of Guelph
Guelph, ON N1G 2W1
uoguelph.ca

University of Lethbridge
Lethbridge, AB T1K 3M4
uleth.ca

University of Manitoba
Winnipeg, MB R3T 2N2
umanitoba.ca

University of Saskatchewan
Saskatoon, SK S7N 5A8
usask.ca

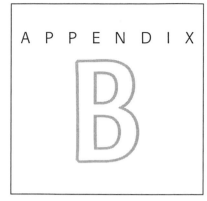

APPENDIX B

PROFESSIONAL ASSOCIATIONS IN HORTICULTURE AND BOTANY

American Association of
  Botanical Gardens and
  Arboreta (AABGA)
100 West 10th Street, Suite 614
Wilmington, DE 19801
aabga.org

American Farm Bureau
600 Maryland Avenue SW,
  Suite 800
Washington, DC 20024
fb.org

American Farmland Trust
1200 18th Street NW
Washington, DC 20036
farmland.org

American Horticultural Therapy
  Association
3570 East 12th Avenue,
  Suite 206
Denver, CO 80206
ahta.org

American Institute of Biological
  Sciences
1444 I Street NW, Suite 200
Washington, DC 20005
aibs.org

American Institute of Floral
  Designers
720 Light Street
Baltimore, MD 21230
aifd.org

American Nursery and
  Landscape Association
1000 Vermont Avenue NW,
  Suite 300
Washington, DC 20005-4914
anla.org

American Phytopathological
  Society
3340 Pilot Knob Road
St. Paul, MN 55121
apsnet.org

American Society for
  Horticultural Science
13 South West Street
Alexandria, VA 22314
ashs.org

American Society of Agricultural
  and Biological Engineers
2950 Niles Road
St. Joseph, MI 49085
asabe.org

American Society of Consulting
  Arborists
15245 Shady Grove Road,
  Suite 130
Rockville, MD 20850
asca-consultants.org

American Society of Farm
  Managers and Rural
  Appraisers
950 South Cherry Street,
  Suite 508
Denver, CO 80246-2664
asfmra.org

American Society of Landscape
  Architects
636 Eye Street NW
Washington, DC 20001-3736
asla.org

American Society of Plant
  Biologists
15501 Monona Drive
Rockville, MD 20855-2768
aspb.org

American Society of Plant
  Taxonomists
Department of Botany
University of Wyoming
Laramie, WY 82071-3165
aspt.net

Association of Professional
  Landscape Designers
1924 North Second Street
Harrisburg, PA 17102
apld.com

Botanical Society of America
P.O. Box 229
St. Louis, MO 63166-0299
botany.org

Cooperative State Research,
  Education, and Extension
  Service (CREES)
U.S. Department of Agriculture
1400 Independence Avenue, Stop
  2201
Washington, DC 20250-2201
csrees.usda.gov

Council of Landscape
  Architectural Registration
  Boards
144 Church Street NW, Suite 201
Vienna, VA 22108
clarb.org

Crop Science Society of America
677 South Segoe Road
Madison, WI 53711
crops.org

Ecological Society of America
1707 H Street NW, Suite 400
Washington, DC 20006
esa.org

Friends of Horticultural Therapy
American Horticultural Therapy
  Association
909 York Street
Denver, CO 80206
hort.vt.edu/human/friendht.html

Golf Course Superintendents
  Association of America
1421 Research Park Drive
Lawrence, KS 66049-3859
gcsaa.org

Institute for Food Technologists
525 West Van Buren, Suite 1000
Chicago, IL 60607
ift.org/cms/

International Society of
  Arboriculture
1400 West Anthony Drive
Champaign, IL 61821
isa-arbor.com

Lady Bird Johnson Wildflower
  Center
4801 La Crosse Avenue
Austin, TX 78739
wildflower.org

Mycological Society of America
P.O. Box 7065
Lawrence, KS 66044
msafungi.org

National Agricultural Library
Abraham Lincoln Building
10301 Baltimore Avenue
Beltsville, MD 20705-2351
nal.usda.gov

National Arbor Day
  Foundation/Institute
100 Arbor Avenue
Nebraska City, NE 68410
arborday.org

National Association of State
  Departments of Agriculture
1156 15th Street NW, Suite 1020
Washington, DC 20005
nasda.org

National Farmers Union
400 North Capitol Street, NW,
  Suite 790
Washington, DC 20001
nfu.org

National Future Farmers of
  America Organization
P.O. Box 68960
6060 FFA Drive
Indianapolis, IN 46268-0960
ffa.org

National Gardening Association
1100 Dorset Street
South Burlington, VT 05403
garden.org

Northeast Organic Farming
Association (NOFA)
P.O. Box 135
Stevens, CT 06491
nofa.org

Organization of Biological Field
Stations
obfs.org

Phycological Society of America
psaalgae.org

Professional Grounds
Management Society
720 Light Street
Baltimore, MD 21230-3816
pgms.org

Professional Landcare Network
(PLANET)
950 Herndon Parkway, Suite 450
Herndon, VA 20170
landcarenetwork.org

Sea Education Association
P.O. Box 6
Woods Hole, MA 02543
sea.edu

Society of American Florists
1601 Duke Street
Alexandria, VA 22314
aboutflowers.com

Tree Care Industry Association
3 Perimeter Road, Unit 1
Manchester, NH 03103
natlarb.com

Wild Farm Alliance
P.O. Box 2570
Watsonville, CA 95077
http://wildfarmalliance.org/

STATE NURSERY ASSOCIATIONS

Alabama
Alabama Nurserymen's
Association
P.O. Box 9
Auburn, AL 36831-0009
alnla.org

Arizona
Arizona Nursery Association
1430 West Broadway, Suite 110
Tempe, AZ 85282
azna.org

Arkansas
Arkansas Green Industry
Association
P.O. Box 21715
Little Rock, AR 72221-1715
No web site

California
California Association of
Nurseries and Garden Centers
3947 Lennane Drive, Suite 150
Sacramento, CA 95834
cangc.org

Colorado

Colorado Nursery and
 Greenhouse Association
959 South Kipling Parkway,
 Suite 200
Lakewood, CO 80226
colorado.nga.org

Connecticut

Connecticut Nursery and
 Landscape Association
P.O. Box 414
Botsford, CT 06404
flowersplantsinct.com

Delaware

Delaware Nursery and Landscape
 Association
P.O. Box 897
Hockessin, DE 19707-0897
No web site

Florida

Florida Nursery, Growers, and
 Landscape Association
1533 Park Center Drive
Orlando, FL 32835-5705
fnga.org

Georgia

Georgia Green Industry
 Association
P.O. Box 369
Epworth, GA 30541
ggia.org

Hawaii

Hawaii Association of
 Nurserymen
Landscape Industry Council
 of Hawaii
P.O. Box 22938
Honolulu, HI 96823-2938
lichawaii.com/Associations/
 han.htm

Idaho

Idaho Nursery and Landscape
 Association
P.O. Box 2065
Idaho Falls, ID 83403
inlagrow.org

Illinois

Illinois State Nurserymen's
 Association
1717 South 5th Street
Springfield, IL 62703
ina-online.org

Indiana

Indiana Nursery and Landscape
 Association
6533 Margaret Court
Indianapolis, IN 46237
inla1.org

Iowa

Iowa Nursery and Landscape
 Association
1210 Frederick Avenue
St. Joseph, MO 64501
iowanla.org

Kansas

Kansas Association of
Nurserymen
411 Poplar
Wamego KS 66547-1446
No web site

Kentucky

Kentucky Nursery and Landscape
Association
350 Village Drive
Frankfort, KY 40601
knla.org

Louisiana

Louisiana Nursery and
Landscape Association
P.O. Box 25100
Baton Rouge, LA 70894-5100
lnla.org

Maine

Maine Landscape and Nursery
Association
500 North Parish Road
Turner, ME 04282
melna.org

Maryland

Maryland Nursery and
Landscape Association
P.O. Box 726
Brooklandville, MD 21022
mnlaonline.org

Massachusetts

Massachusetts Nursery and
Landscape Association
P.O. Box 387
Conway, MA 01341-9772
mnla.com

Michigan

Michigan Nurserymen and
Landscape Association
2149 Commons Parkway
Okemos, MI 48864
mnla.org

Minnesota

Minnesota Nursery and
Landscape Association
http://gardenminnesota.com

Mississippi

Mississippi Nursery and
Landscape Association
msnla.org

Missouri

Missouri Landscape and Nursery
Association
mlna.org

Montana

Montana Nursery and Landscape
Association
P.O. Box 4553
Missoula, MT 59806-4553
plantingmontana.com

Nebraska

Nebraska Nursery and Landscape
    Association
4200 Witherbee Boulevard
Lincoln, NE 68510
nnla.org

Nevada

Nevada Landscape Association
P.O. Box 7431
Reno, NV 89510-7431
nevadanla.org

New Hampshire

New Hampshire Plant Growers
    Association
25 Riverbend Road
Newmarket, NH 03857
http://nhplantgrowers.org/

New Jersey

New Jersey Nursery and
    Landscape Association
gardennj.net

New Mexico

New Mexico Association of
    Nursery Industries
Box 30003 Dept 3Q
Las Cruces NM 88003-8003

New York

New York State Nursery/
    Landscape Association
2115 Downer State Road
P.O. Box 657
Baldwinsville, NY 13027-9702
No web site

North Carolina

North Carolina Association of
    Nurserymen
968 Trinity Road
Raleigh, NC 27607
ncan.com

North Dakota

North Dakota Nursery and
    Greenhouse Association
P.O. Box 34
Neche, ND 58265
ndnga.com

Ohio

Ohio Nursery and Landscape
    Association
72 Dorchester Square
Westerville, OH 43081-3350
onla.org

Oklahoma

Oklahoma Nursery and
    Association
400 North Portland
Oklahoma City, OK 73107
oknurserymen.org

Oregon

Oregon Association of
    Nurseries
29751 SW Town Center
    Loop W.
Wilsonville, OR 97070
nurseryguide.com

Pennsylvania

Pennsylvania Landscape and
    Nursery Association
1707 South Cameron Street
Harrisburg, PA 17104-3148
plan.com

Rhode Island

Rhode Island Nursery and
    Landscape Association
64 Bittersweet Drive
Seekonk, MA 02771
rinla.com

South Carolina

South Carolina Nursery and
    Landscape Association
332 Sunward Path
Inman, SC 29349
scnla.com

South Dakota

South Dakota Nursery and
    Landscape Association
5659 Dakota South
Huron, SD 57350-6550
No web site

Tennessee

Tennessee Nursery and
    Landscape Association
P.O. Box 57
McMinnville, TN 3 7110
tnla.com

Texas

Texas Nursery and Landscape
    Association
7730 South IH-35
Austin, TX 78745-6698
txnla.org

Utah

Utah Nursery and Landscape
    Association
P.O. Box 526314
Salt Lake City, Utah 84152
utahgreen.org

Vermont

Vermont Association of
    Professional Horticulturists
P.O. Box 396
Jonesville, VT 05466-0396
vaph.org

Virginia

Virginia Nursery and Landscape
    Association
383 Coal Hollow Road
Christiansburg, VA 24073-6721
vnla.org

Washington

Washington State Nursery and
    Landscape Association
34400 Pacific Highway South,
    Suite 2
Federal Way, WA 98003
wsnla.org

West Virginia

West Virginia Nursery and
  Landscape Association
1517 Kingwood Pike
Morgantown, WV 26508
wvnla.org

Wisconsin

Wisconsin Green Industry
  Federation
12342 West Layton Avenue
Greenfield, WI 53228
wislf.org

APPENDIX

C

SUGGESTED READING

The following list represents just some of the many titles available to those interested in the careers discussed in this book.

Acquaah, George. *Horticulture: Principles and Practices*. 3d ed. Upper Saddle River, NJ: Prentice-Hall, 2004.

Adams, C.R., and M.P. Early. *Principles of Horticulture*. 4th ed. Cambridge: Butterworth-Heinemann, 2004.

Bahamon, Alejandro. *Ultimate Landscape Design*. Berlin: te Nueus Publishing Co., 2006.

Camenson, Blythe. *Careers for Plant Lovers and Other Green Thumb Types*. 2d ed. New York: McGraw-Hill, 2004.

Coyle, Heather M. *Forensic Botany: Principles and Applications to Criminal Casework*. Boca Raton, FL: CRC Press, 2004.

de Jong-Stout, Alisa A. *A Master Guide to the Art of Floral Design*. Portland, OR: Timber Press, 2006.

Dole, John M., and Harold F. Wilkins. *Floriculture: Principles and Species*. 2d ed. Upper Saddle River, NJ: Prentice-Hall, 2004.

Gurevitch, Jessica, et al. *The Ecology of Plants*. Sunderland, CT: Sinauer Associates, 2002.

Jozwik, Francis X., and John Gist, eds. *Plants for Profit: Income Opportunities in Horticulture*. Mills, WY: Andmar Press, 2000.

Kimball, Cheryl. *Start Your Own Florist Shop and Five Other Floral Businesses*. New York: Entrepreneur Press, 2006.

Larcher, Walter. *Physiological Plant Ecology*. 4th ed. New York: Springer, 2003.

Lewis, Jerre G., and Leslie D. Renn. *How to Start and Manage a Retail Florist Business: A Practical Way to Start Your Own Business*. Lewis & Renn Associates, Inc., 2004.

Lewis, Walter H., and P.F. Memory Elvin-Lewis. *Medical Botany: Plants Affecting Human Health*. 2d ed. New York: Wiley, 2003.

Mauseth, James D. *Botany: An Introduction to Plant Biology*. 3d ed. Sudbury, MA: Jones and Bartlett, 2003.

Maxted, Nigel. *Plant Genetic Conservation*. Cambridge: Cambridge University Press, 2006.

McIndoe, Andrew. *Design and Planting (Horticulture Gardener's Series)*. Horticulture Books, 2006.

Melby, Pete, and Tom Cathcart. *Regenerative Design Techniques: Practical Applications in Landscape Design*. New York: Wiley, 2002.

Nabors, Murray. *Introduction to Botany*. Upper Saddle River, NJ: Benjamin Cummings, 2003.

Nelson, Paul V. *Greenhouse Operation and Management*. 6th ed. Upper Saddle River, NJ: Prentice-Hall, 2002.

Pastor, Sharon, and Martha C. Straus, eds. *Horticulture as Therapy: Principles and Practice*. Binghamton, NY: Haworth Press, 2003.

Poincelot, Raymond P. *Sustainable Horticulture: Today and Tomorrow*. Upper Saddle River, NJ: Prentice-Hall, 2004.

Rice, Laura W., and Robert P. Rice. *Practical Horticulture*. 6th ed. Upper Saddle River, NJ: Prentice-Hall, 2005.

Rogers, Elizabeth B. *Landscape Design: A Cultural and Architectural History*. New York: Harry N. Abrams, 2001.

Slater, Adrian R. et al. *Plant Biotechnology: The Genetic Manipulation of Plants*. New York: Oxford University Press USA, 2003.

Smith, Ed, and Mickey Bassett. *Classic Floral Designs*. New York: Sterling Publishing, 2005.

Trigiano, Robert M. *Plant Development and Biotechnology*. Boca Raton, FL: CRC, 2004.

ABOUT THE AUTHOR

After ten years as a professor of horticulture and chairman of a horti-culture department in the Virginia Community College System, Dr. Jerry Garner joined the staff of Mariani Landscape in Chicago, Illinois as their Horticulturist and Plant Buyer. In this position, he brought his extensive academic and consulting expertise in residential and commercial plant materials and garden design to one of the Midwest's most prestigious design/build firms specializing in estate work.

In addition to a bachelor's degree in English and graduate work in edu-cation from The College of William and Mary, Dr. Garner holds a bache-lor's degree in horticulture with a specialization in floriculture from Virginia Polytechnic Institute and State University (Virginia Tech). Prior to his bachelor's work in horticulture, Dr. Garner was employed as a hor-ticultural therapist and was extensively involved in the professional asso-ciation in this area. After receiving his bachelor's degree, he pursued a Ph.D. in horticulture at Virginia Tech.

In Virginia, Dr. Garner served as the horticultural advisor to major busi-nesses, government agencies, public universities, and industry executives. He also served on numerous horticultural advisory boards, including the Virginia State Fairground Horticulture Pavilion and the Maymont Foundation and Park. In addition, he was a regular lecturer at the Lewis Ginter Botanical Gardens.

Dr. Garner was a founding board member of the Maymont Flower and Garden Show in Richmond, Virginia, and served on the executive board of the Virginia Society of Landscape Designers. Since coming to Illinois, Dr. Garner has become an active member of the Illinois Landscape Contractors Association, serving on the Education Seminars Committee. He is also on the Board of Directors of the Evanston Arts Center where he serves on the Exhibition, Benefits, and Nominating Committees.

This edition was revised by Josephine Scanlon, a freelance writer who specializes in career guides.